PUTT'S LAW

&

THE SUCCESSFUL TECHNOCRAT

PUTT'S LAW

&

THE SUCCESSFUL TECHNOCRAT

How to Win in the Information Age

Archibald Putt

Illustrated by Dennis H. Driscoll

IEEE PRESS

WILEY-INTERSCIENCE

A JOHN WILEY & SONS, INC., PUBLICATION

Published by John Wiley & Sons, Inc., Hoboken, New Jersey.
Published simultaneously in Canada.

For general information on our other products and services or for technical support, please contact our Customer Care Department within the United States at (800) 762-2974, outside the United States at (317) 572-3993 or fax (317) 572-4002.

Wiley also publishes its books in a variety of electronic formats. Some content that appears in print may not be available in electronic format. For information about Wiley products, visit our web site at www.wiley.com.

Library of Congress Cataloging-in-Publication Data is available.

ISBN-13 978-0-471-71422-4
ISBN-10 0-471-71422-4

Printed in the United States of America.

10 9 8 7 6 5 4 3 2 1

Author's Note

This book is about people and organizations that work with modern technology. The people, organizations, and events are fictitious—except in obvious cases when they are real. The factors that affect them and the pressures that drive them are always real.

Contents

Preface

When the first edition of *Putt's Law and the Successful Technocrat* was published, I often enjoyed finding myself in a group of engineers who were discussing the book. Because it was published under the pseudonym, Archibald Putt, they did not know that I was the author. Being a celebrity is fun. Being an unknown celebrity is, well, a different type of fun.

A number of colleagues who know my secret identity have urged me to publish a new edition, noting that thousands of references to Putt's Law can be found on the Internet, some in the Dutch, Finnish, and German literature, often with the notation, "author unknown." This evidence of continuing worldwide interest helped convince me that a new generation of technocrats (and those who must work with them) would enjoy and benefit from my book.

I was also encouraged by the dearth of superior advice for ambitious technocrats in more recent literature. For example, *The Dilbert Principle* by Scott Adams provides wonderful insights and chuckles for nonmanagers but little guidance for ambitious employees who want to advance in management.

Revising my book was gratifying. All of the original laws and corollaries are still valid, so I changed none, but I did add many new ones. The most significant additions relate to advances in information technologies that have changed forever the way people work and interact with each other. New analyses, first revealed in this edition, will be valuable to all who aspire to win in the Information Age.

I have also introduced my recently developed Method of Rational Exuberance, which practically guarantees a rapid rise in management. And I have answered the often asked question, "Can Putt's Law be broken?"

Some scholarly types have suggested that my writings should be viewed merely as humorous satire. Holding that view can inhibit your success. It is not the view of many successful technocrats who studied and used the lessons of my book. While winning the game, they laughed just as often as others, especially on the way to the bank.

Some readers may occasionally be confronted by unfamiliar technical terms and scientific analyses, which are essential to the acceptance of the book by the technical community. If such a passage confounds you, simply reread it with an air of confidence or boredom while occasionally muttering a knowing "uh-huh." With practice you will find your comprehension is greatly improved, and you will soon have mastered one of the most important techniques of the successful technocrat.

ARCHIBALD PUTT

Part One

PUTT'S PRIMER

Putt's Law and Corollary

Years of study of the sociology of organizations that develop or work primarily with sophisticated new technologies have convinced me that such organizations are quite different from those in other fields. Thus, the excellent P-literature by Parkinson, Peter, and Potter, which describes so clearly most social hierarchies, is inadequate and misleading when applied to fields of high technology.

The Peter Principle, for example, states that "in a hierarchy every employee tends to rise to his level of incompetence."* We are all familiar with the workings of this principle in typical organizations. Employees are eligible for promotion until they reach a level in which they can no longer perform effectively. They then have reached their level of incompetence and are no longer promotable.

Upon reflection, the Peter Principle seems to be self-evident. Why, then, did it take so long to be recognized? Its corollary, "In time, every post tends to be occupied by an employee who is incompetent to carry out its duties," explains why even simple things are so often bungled in large organizations.

It is virtually impossible for a person to avoid reaching a level of incompetence in a hierarchy. Once a promotion has

---

*From *The Peter Principle*, by Laurence J. Peter and Raymond Hull, Morrow, NY, 1969.

been offered, ego conspires with social pressure to force its acceptance. The hapless engineer who turns down a managerial job because he or she prefers working as an engineer can only expect problems. How can they explain this to their family? And how can their spouse explain this lack of ambition to parents or to friends? The engineer may refuse the first offer of a promotion, but is unlikely to refuse the next. Thus begins the inexorable rise to his or her level of incompetence.

CREATIVE INCOMPETENCE

The only satisfactory way to avoid reaching one's level of incompetence, according to Peter, is through *creative incompetence.* This is achieved by developing a high level of incompetence in some area that does not affect present performance, but does assure that there will be no further offers of promotion. This is an uncommon tactic in typical hierarchies.

It may readily be observed, however, that creative incompetence is the rule rather than the exception in hierarchies that work with science and technology. Consider, for example, the case of Bob Bottomly who had been employed in the development laboratory of a large electronics firm for several years. He learned that his superior was to be promoted and that he was the most likely candidate for the vacated position. The next day he took off his shirt in the laboratory and continued to work in his undershirt, complaining loudly about the heat. When his superior was promoted, it was another member of the group who was chosen to fill the vacancy. Bob Bottomly soon returned to wearing a shirt, except on the rare occasions when he felt promotions might be under consideration.

Then there is the case of Dr. Schwartz, whose phenomenal grasp of the literature in physical chemistry made her a great asset in the central research laboratory of a major corporation. She was able to carry hundreds of references to the literature in her head and had many thousands carefully organized in her computer. A query to her about almost any subject in this field would produce more information in a few minutes than could be obtained by searching for hours on the Internet or for many days in a reference library.

Dr. Schwartz was under consideration to become the manager of the chemistry group until it was learned that she continually misplaced administrative records and failed to attend management briefings. Such traits were unacceptable for a manager, and so Dr. Schwartz continued for years doing her own research and keeping a well-organized file of chemical literature as a service for herself and others in the laboratory.

Perhaps the best-known case is that of Albert Einstein, the preeminent scientist of the twentieth century. In a time when long hair was not common, his was long and bushy. He typically wore an open-collared shirt, an old sweater, baggy trousers, and no socks. Thus, he seldom had to contemplate major administrative jobs and spent his time in positions where he could concentrate on theoretical physics.

Such examples of creative incompetence are so numerous in science and technology that many low-level positions remain staffed by competent people who never reach their level of incompetence. These people find their satisfaction in technical work and would be bored and frustrated by administrative responsibility.

Successful technocrats, however, are not found among the ranks of plodders of such limited ambition. Instead, they are found among those who aspire to eminence through their *position* in the technical hierarchy. Such people will find their climb made easier by the many individuals who choose to remain behind by practicing creative incompetence.

NO CRITERION FOR COMPETENCE

If the large number of people practicing creative incompetence were the only anomaly in technological hierarchies, we might conclude that persons aspiring to higher placement would be promoted to their level of incompetence, as is typical in most hierarchies. However, there is yet another anomaly with most interesting consequences: there frequently is no way to judge whether an individual is competent to hold a given position. Stated another way, there is no adequate *competence criterion* for technical managers.

Consider, for example, the first U.S. space laboratory. When it was placed into orbit in May 1973, its meteoroid and thermal shielding and one of its solar cell wings were torn away. Another solar cell wing was jammed closed so that only two of the wings deployed properly. This threatened the mission with failure. Should the project leader have been fired for failure to prevent this problem or should he have been given a citation for erecting a makeshift parasol to shield the laboratory from the sun and getting the jammed wing deployed during a space walk by the astronauts?

Then there is the case of a small research group in the Pfizer Central Research Laboratory in Sandwich, England. Beginning in 1986, their leader set the objective of developing a drug for treating hypertension. Two years later, he changed the primary objective to the treatment of angina pectoris. Progress was slow. In 1992, clinical trials were undertaken to test patient-toleration levels for the compound known as UK-92,480. Among the side effects observed at different dosage levels were indigestion, back aches, and leg pains.

"Oh, there are also some reports of penile erections," one clinician observed. But this observation was not thought to be particularly significant. Thus, it was not until 1994 that clinical trials were undertaken of compound UK-92,480 (sildenafil citrate) for erectile dysfunction. The results were excellent, but a market analysis indicated that sales would be small.

In March 1998, U.S. FDA approval was given for the sale of sildenafil citrate for treating erectile dysfunction. Given the name Viagra, three million prescriptions were written during its first three months in the market—dramatically more than Pfizer anticipated. Sales rose rapidly, grossing Pfizer $1.7 billion during its fifth year in the market. Viagra had become a blockbuster drug, providing Pfizer with unprecedented publicity and profits.

Should the researchers have been chastised for failing to accomplish their original objectives, or should they have been praised for finding an unexpected use for their compound? Should someone have been fired for responding so slowly when clinical tests revealed penile erections to be a "side effect"? And what about the marketing study that grossly underestimated the demand for such a drug?

In an advanced research or development project, success or failure is largely determined by the goals and objectives set before a manager is chosen. Although a hardworking and diligent manager can increase the chances of success, the outcome of the project is most strongly affected by preexisting, but unknown, technological and societal factors. Sometimes, a manager can save a project from failure by changing its objective. But the likelihood of success is still more a matter of luck than planning. The success or failure of a project should, therefore, not be used as the primary measure of a manager's competence.

PUTT'S LAW IS PROMULGATED

Without an adequate competence criterion for technical managers, there is no way to determine when a person has reached his or her level of incompetence. Thus, a clever and ambitious person may be promoted from one level of incompetence to another. This phenomenon, combined with the practice of creative incompetence by those who understand technology best, provides the basis for Putt's Law, which can be stated in an intuitive and nonmathematical form:

> Technology is dominated by two types of people:
> those who understand what they do not manage
> and those who manage what they do not understand.

Just as in any other hierarchy, the majority of people neither understand nor manage much of anything. This does not create an exception to Putt's Law because such persons do not dominate the hierarchy.

At the same time that many outstanding technical people choose to stay near the bottom of the hierarchy, ambitious individuals with less understanding of technology will be promoted again and again until one of them reaches the very top. This ultimate result of Putt's Law is well described by Putt's Corollary, which is more formally known as the First Corollary to Putt's Law:

> Every technical hierarchy, in time,
> develops a competence inversion.

Competence inversion in a technical hierarchy.

Putt's Law and Putt's Corollary are axiomatic to organizations that develop or make extensive use of sophisticated technologies. Indeed, success in these organizations depends upon one's ability to deal with and benefit from these fundamental attributes. Whether competent or not, technocrats *can* rise in management by knowing and obeying the Laws and Rules of the Hierarchy.

The pronouncement of Putt's Law and its Corollary brought an instant and enthusiastic response. In an early private poll, 94 percent of all technologists responded with a resounding affirmation of Putt's Law and Putt's Corollary. Only six percent (all managers) disagreed.

In raising their objections, these managers said it was not possible that so many technological advances had been achieved by organizations with competence inversions. Often, they pointed with pride to the contributions of their own organizations. But in doing so they showed that they had missed the point. Much of the success of technology can be attributed directly to the tendency for competent technologists to remain near the bottom of organizational hierarchies.

In nontechnical organizations, key positions remain blocked once they are filled by incompetent people. In a technical hierarchy, by contrast, incompetent individuals continue to rise. Incompetence is thus flushed out of the lower levels, leaving competent people behind to do the work. In fact, Putt's Law and Putt's Corollary can be regarded as prime reasons for continued advances in technology.

AN EXAMPLE IN CONTRAST

The medical profession is an interesting example in contrast, for it has virtually no hierarchy. Most doctors, competent or not, continue to treat patients throughout their careers. They cannot be promoted into less critical positions because so few administrative positions exist. Each year, numerous medical cases are reported that are suggestive of this problem.

In one instance, a young hospital patient was given a local anesthetic prior to a simple tonsillectomy. The anesthetic was accidentally injected into the carotid artery leading directly to the brain. The patient died almost immediately. In another case, a middle-aged man entered a leading hospital for replacement of a damaged heart valve. When he was taken off the heart–lung machine following the operation, he died within minutes. An autopsy revealed that the surgeon had put the artificial valve in backward.

In yet another case, doctors at a major university hospital gave a 17-year-old patient a heart and lung transplant from a donor with an incompatible blood type. "Severe and irreversible brain damage" resulted when the doctors attempted to correct the error by implanting replacement organs of the correct blood type.

While similar things can, and do, occur in large technical hierarchies, incompetent technical people experience a social pressure from their more competent colleagues that causes them to seek security within the ranks of management. In technical hierarchies, there is always the possibility that incompetence will be rewarded by promotion.

Two Laws of Crises

T he preceding chapter dealt with two anomalies of the technical hierarchy that provide a major basis for Putt's Law. However, there is yet another important anomaly: *A person must rock the boat to get ahead.* This is just the reverse of typical hierarchies, in which rocking the boat is unacceptable and definitely not conducive to promotions.

This third anomaly results from the fact that rapid progress in technology is always accompanied by great uncertainty. In *blue-sky projects,* for example, the problems and benefits are unpredictable, but both are expected to be large. In such projects, it is expected that target dates will be missed and that additional people and equipment will be required. *Pushing the state of the art* is another popular phrase that is used to describe projects designed to advance technology more rapidly than has occurred in the past. Such projects invariably lead to problems and crises. In fact, if there were no crises, it would be presumed that the goals were not aggressive enough—and no technologist can afford to have projects with insufficiently aggressive goals.

Thus, the importance of rocking the boat or of having imperfections and crises in the projects of a technical hierarchy are evident. The First Law of Crises follows rather logically from these observations and may be stated as follows:

Technical hierarchies abhor perfection.

Putt's Law and the Successful Technocrat. By Archibald Putt
Copyright © 2006 the Institute of Electrical and Electronics Engineers

A person must rock the boat to get ahead.

The implication of this to ambitious technologists is clear—they must avoid perfection. This admonition is unnecessary for most persons, who could not achieve perfection even if they tried. However, it will be of help to some. Consider, for example, the case of Roger Proofsworthy, whose excessive competence caused him to labor needlessly at the bottom of the hierarchy for many years.

Proofsworthy was hired into the development laboratories of the Ultima Corporation shortly after receiving his Ph.D. in electrical engineering. He was a man who combined a solid academic background and good technical insight with a dogged refusal to be less than perfect in all his activities. Once he was given an assignment, it was as good as done.

In his second year with the company, he was placed in charge of the transfer of a technical innovation from the development laboratory to the manufacturing division. As usual, his performance was outstanding. When technical problems were uncovered, he solved them, often working around the clock until the job was done. At times, he found himself embroiled in the difficult political problems associated with the reluctance of the manufacturing division to accept the work done in the development laboratory. However, he managed to resolve all of these difficulties before they became major issues in the corporation. Within a few months, he had accomplished a task that normally would have taken years and involved persons at the highest levels of the corporation.

Despite this outstanding performance, Proofsworthy did not receive the next promotion. It went instead to a colleague of substantially less capability who had difficulty handling his own projects. These difficulties often went unsolved until corporate officers became involved. As a result, the colleague became known throughout the corporation as an individual who handled difficult assignments. He was found to be personable in his interactions with management and quite levelheaded. He was thus a logical choice for promotion.

Proofsworthy, in contrast, was unknown beyond his immediate manager, and even his own manager had little reason to know of the difficult problems Proofsworthy had personally handled.

The colleague chosen for the promotion in this case went on to achieve further promotions and rapidly reached a high level in the hierarchy. Proofsworthy remained a staff engineer for many years. He finally left Ultima in disgust. Unfortunately, the problem that plagued him at Ultima continued to plague him elsewhere. He always achieved such a high level of perfection that his accomplishments went unnoticed.

PERFECTION BRINGS NO REWARD

Such perfection as described above is seldom seen near the top of a technical hierarchy because individuals so afflicted are usually unable to progress beyond the first or second levels. Nevertheless, my extensive research has found one such case in a very small company.

Cosmo J. Draper was hired as director of research and development for a small company whose major product line was becoming technically obsolete. Within a year, he put together a cadre of creative people who saved the line from obsolescence and went on to assure the company's technological leadership in this area.

As the years passed by, Draper continued to strengthen the research activities, expanding their scope to include all fields of interest to the company. He also established effective working relationships with the product groups so that innovations moved smoothly from research to development and then into the manufacturing division. As a result, he became a member of the management team.

Now that the members of top management no longer had cause to become involved in technological issues, they spent most of their time on marketing and financial problems. The memory of the crisis that caused them to hire Draper gradually faded. Finally, in an economy move, they gave Draper early retirement and dismissed four of his top technologists and some support personnel. What remained of Draper's organization was absorbed into the manufacturing groups. The chairman of the board was pleased to advise stockholders at the company's annual meeting of the belt-tightening measures that had resulted in a small increase in profits.

These examples of the hazards of excessive perfection clearly confirm the validity of the First Law of Crises. They further suggest that any manager who is competent enough to avoid crises entirely should nevertheless introduce some into his operation. A manager could, for example, deliberately fail to meet an objective set by top management. This would assure his survival and upward progress. In this way, even an exceptionally competent person can rise as rapidly as an individual with just the *right* amount of incompetence for the job.

But what is the right amount of incompetence or the right amount of crisis to introduce into a given job? The answer to that question is partially answered by the Second Law of Crises:

> The maximum rate of promotion is achieved
> at a level of crisis only slightly less
> than that which results in dismissal.

Because of the precarious nature of the boundary between cause for dismissal and the maximum rate of promotion, a prudently ambitious technocrat should not begin a new assignment with too high a level of crisis. Care is especially recommended because the exact level of crisis permitted before dismissal may not be well understood until after some time in a new assignment. Furthermore, some crises may occur spontaneously and unavoidably.

FIXING THE INCOMPETENCE LEVEL

The best strategy is to begin a new assignment with as low a level of crisis as possible. The level should then be increased gradually until the desired promotion occurs. The optimum timing is illustrated in the figure on crisis level strategy on the following page.

The solid line shows how an astute technocrat gradually increases the level of crisis from zero up toward the dismissal level, which is represented by the broken line in the figure. He judges when his next promotion should occur and attempts to achieve 63% of the dismissal level by that time (t_0 in the figure). After twice that time ($2t_0$), his level of crisis will have reached 86% of the dismissal level.

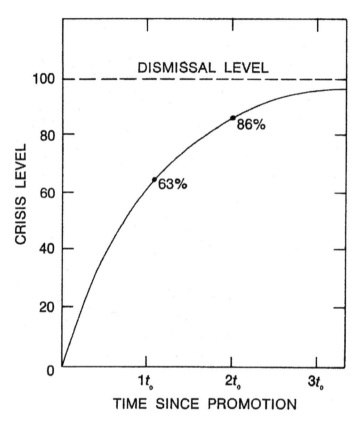

Crisis level strategy.

If still no promotion results, he continues to increase the amount of crisis toward the dismissal level. Simultaneously, he should begin looking for another job because any company that fails to respond by the time 86% of the maximum permitted level of crisis has been reached is no place for an ambitious technocrat.

A MATHEMATICALLY PRECISE STRATEGY

Mathematically trained readers will be pleased to know that this strategy can be represented even more precisely by the equation

$$C = C_0(1 - e^{-t/t_0})$$

where C is the actual level of crisis, C_0 is the level of crisis that results in dismissal, t is the time since the last promotion, t_0 is the expected time between promotions, and e is a universal constant having an approximate value of 2.7182818285.

Readers with limited mathematical training should try reading the equation and explanation of terms out loud, while replacing each comma with a thoughtful "uh-huh" and the final period with the most confident sounding "uh-huh" they can muster. This greatly improves one's comprehension and is one of the most important techniques of successful technocrats.

The Law of Failure
and Putt's Ploy

We have now seen how a person can rise in the hierarchy by having the right amount of incompetence or by artificially adding crises to compensate for his own excessive perfection. But what about the person with more than the desired amount of natural imperfections? Can a person succeed if his own incompetence prevents him from staying below the maximum tolerable crisis level?

Fortunately for most, the answer is yes. There is hope of promotion even for the truly incompetent. However, a proper appreciation is needed of the Law of Failure:

> Innovative organizations abhor little failures
> but reward big ones.

Consider, for example, the problem faced by two corporate officers trying to select a manager for an important development project. The first person under consideration managed three minor projects in the past and each one failed. Such a person is not the right choice. The second person had successfully managed several small projects and is thus a good candidate. The third person had managed a project that became one of the largest technical failures in the history of the company. As a result, he had greater technical management experience than ei-

**Innovative organizations abhor little failures
but reward big ones.**

ther of the other candidates and, no doubt, had learned a lot from the failure.

The corporate officers had the wisdom to know that failure in high-technology projects is frequently unavoidable. So, after due consideration, they selected the third candidate to manage the new project.

Seems unlikely? Not at all. When management is forced to choose between a person with demonstrated successes in small projects and one with a demonstrated failure in a large project, it more often than not opts for the person with a large failure. From this reality is derived the Corollary to the Law of Failure, which is normally stated in its alliterated form:

Failure to fail fully is a fool's folly.

What an ambitious person should do when faced with failure is made explicit by Putt's Ploy:

If you must fail, fail big.

Even in the raw form of this brief statement, Putt's Ploy can be useful. But much greater benefits can be achieved by knowing and using the optimum timing for failures, as revealed in the figure on the following page. Curve A is the optimum strategy for introducing crises, which was discussed in the previous chapter. Curve B is the graphical representation of Putt's Ploy. It shows what to do if you find yourself unavoidably too far above curve A.

If one's level of crisis exceeds 63% of the dismissal level well before the time t_0 (when the earliest promotion might be expected), quick action is needed. New crises should be introduced as rapidly as possible in order to pass quickly through the forbidden zone of minor failure. Every effort should be made to exceed the major-failure level in a time less than $2t_0$. Once a person is "safely" above the major-failure level, his job is once again secure. Management will be unable to find anyone "qualified" to take over such a project.

It is generally best to stop increasing the level of crisis before the company becomes bankrupt. However, there is a more advanced ploy in which the company is driven into bankrupt-

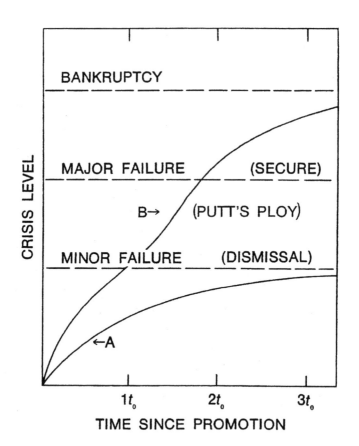

Putt's Ploy.

cy. In the simple version presented here, the manager holds the level of crisis above the major-failure level and under the bankruptcy level, until he and top management solve the problem. An important aspect of the solution must of course be a satisfactory new assignment for the troubled technocrat.

Some scholarly persons have suggested that a technocrat should not waste his time trying to follow the controlled introduction of crises of curve A. Rather, he should immediately target a course to get above the major-failure level as quickly as possible. This suggestion is not consistent with Putt's Ploy, however, and fails to take proper account of the uncertainties at each step.

My analysis shows that a person who follows curve A has a probability of being promoted before time $2t_0$ of 97.3%. Thus, if one assumes that ten promotions will be required to reach the top, the probability of getting there through ten consecutive promotions is quite good, namely, 76.1%.

On the other hand, there is no certainty that every big failure will be followed by a big promotion. My studies show that the probability is only slightly better than fifty-fifty, that is, 57.4%. Assuming that five *big* promotions are needed to get to the top, the probability of getting there through five *big* failures is thus 6.2%, which is only about twice as good as the probability, in tossing a coin, of turning up heads five times in a row.

Most people would not be willing to risk their careers on odds like this. But for those of virtually no technical competence at all, the path of big failures does provide a 6.2% chance of reaching the very top!

Three Laws of Innovation

Large failures may lead to large promotions, but small failures never do. They lead only to oblivion. Thus, it is important to convert small failures into large failures by using Putt's Ploy as discussed in the previous chapter.

But what if a small failure cannot be turned into a big one? Is there some way to avoid the addition of another small failure to one's growing list?

Fortunately, there is a way. The troubled technocrat must convince his management that the failure is actually a success. General guidelines do not exist, so there is considerable opportunity for innovation. In fact, some of the most innovative thinking in technology has been applied to converting technical failures into personal successes. Successes so achieved are referred to as *innovated successes.*

Consider, for example, the professor of chemistry who spent many years in a fruitless effort to separate two isotopes of a rare metal. While unsuccessful at his project, he nevertheless traveled to several conferences to present his findings. Finally, he wrote a book describing the problems of isotope separation. This volume is now prominently displayed on his desk, providing clear evidence of a successful career to which other professors may aspire.

Then there is the case of the aerospace firm that contracted to develop a new weapon system. After many schedule slippages and cost overruns, the contract was terminated without completing the weapon system.

This might have proved embarrassing to the firm and to the military if the final report had not found so many contributions to highlight. There was advancement of the state-of-the-art, establishment of technical limits, development of new concepts of potential value for future systems, and countless areas of technical fallout. What at first appeared to be a failure was actually a success when viewed with the "broader, long-range perspective" of the final report.

These leaders of technology had verified once again the First Law of Innovation:

> An innovated success is as good
> as a successful innovation.

Many situations require more sophisticated schemes. Consider, for example, the problem faced by George Clearwater, Director of Environmental Engineering at the Ultima Corporation. A junior staff engineer in his department named Roger Proofsworthy had pointed out a major flaw in one of the development projects only days before it was to be transferred to manufacturing. Now there was no hope of meeting, or even coming close to, the schedule. A solution could not be expected quickly, if ever.

How should Clearwater advise management? Should he place all the blame on the project leader—a man he himself had hired? The error pointed out by the staff engineer seemed so obvious. How could it have happened? The more he thought about it, the more he fumed. He picked up his phone and made an appointment with the project leader.

A FATEFUL MEETING

The project leader showed up promptly. He knew the mood of his boss and decided to take the initiative quickly.

"George, I've been thinking about this pollution control project," said Ralph Striker, the project leader, as he initiated the discussion. "We've run into a bit of a snag, but the problem

Proofsworthy uncovered wouldn't normally have been found until after the manufacturing engineers got it. It's the type of mechanical problem they deal with all the time, so I think we should transfer it to them as is."

"And not tell them about the problem? We couldn't get away with that, Ralph—not with Proofsworthy knowing what he does."

"How about Research," Ralph suggested. "We could transfer the whole thing to them and suggest it only needs a few basic studies. They'd fool around with it a long time before spotting this problem."

"You're right there," responded George with enthusiasm, "but they'll never take it. Too much NIH."

"Yeah, no group has more of the *not invented here* syndrome than they do."

"If Proofsworthy was from Research, they might take the project just 'cause one of their own people discovered the problem."

"Yeah," said Ralph, "they're dumb enough. Besides, they could write a bunch of papers for some technical journal."

"And give out a few achievement awards before terminating the project," added George.

They both chuckled. One thing they could always agree on was their attitude toward the geeks and nerds in Research.

ON THE RIGHT TRACK

"I think," said George, after more reflection, "that you're on the right track. The project should be transferred as is."

"No group'll wanta try to solve that mechanical problem once they talk to Proofsworthy," responded Ralph with evident concern.

"Probably, but wasn't there some chemical problem, too?"

"Yeah, but it's not serious. We practically have it solved."

"Well," said George, "I think the solution is clear. Have your people write up the chemical problem. Define it, but don't solve it. Then ask the Chemical Engineering Department to take responsibility for the project. Tell 'em it needs their type of expertise. They'll like that."

"But what about Proofsworthy? If he points out the really big problem, then what?"

"No sweat there. Just let it be known that you aren't satisfied with his performance, and I'll transfer him to another project in my department. Then if he raises a fuss about the mechanical problem, people'll think it's a case of sour grapes. I'll see to it that he doesn't give you any problems."

"Okay," responded Ralph with appreciation. Then after a bit of thought, he added, "I think I should recommend an outstanding contribution award to the guy who spotted the chemical problem. That should keep attention focused where we want it."

"Sounds good," said George. "I'll see to it that the corporate officers approve the award. In addition, I'll recommend a raise for you. With that kind of picture, the Chemical Engineering Department will beg us for the project."

A STROKE OF GENIUS

George Clearwater and his trusted subordinate carried out the decisions of their meeting with a firm dispatch befitting their managerial positions. The project was successfully transferred. Awards and rewards were liberally distributed. Through *innovative* management methods, they had converted a technical failure into an *innovated success.* Like so many managers before them, Clearwater and Striker had learned the meaning of the First Law of Innovation.

However, the real stroke of genius was recommending the raise for Ralph. We can only hope it was a big one, for, as pointed out by the Second Law of Innovation:

> The true measure of success
> in an innovative project
> is the size of management's reward.

Roger Proofsworthy paid little attention to these events. He was much too busy on another project, which was in serious difficulty before he arrived. Proofsworthy was made to understand how lucky he was to have another chance to "prove himself." His progress would be closely watched.

THE INNOVATIVE PROCESS

Striker and Clearwater carried out their assignments well. The Environmental Engineering Group had worked on the pollution control problem with all its skills and resources. Some problems were solved and others identified. Outside help was sought when needed, and, finally, arrangements were made for transferring the project to another group for more development effort.

This is the nature of the innovative process in large organizations. No one person or even one group can carry an innovation from conception through to manufacturing and sales. There are too many specialists, each ready to play a role, each insistent on playing that role.

Thus, at any stage in the development process, there must be negotiations among groups. Coordinated programs are essential, and so are frequent transfers of projects from one group to another. Even if a project does not require a variety of technical skills, it still must pass from group to group as it makes its way from basic research to applied research, advanced development, product development, and, finally, manufacturing and sales. As stated in the Third Law of Innovation:

> Innovation may be the goal,
> but technology transfer is the business
> of technical hierarchies.

Books have been written about it. Seminars have been devoted to it. But technology transfer remains an art. Each transfer is the result of negotiations that might well tax the skill of a seasoned diplomat. It can only occur if both sides perceive a benefit or if a higher authority demands it. Because the real reason for a proposed transfer of technology may not be known to the receiving group, there is often considerable resistance.

Technical innovations *should* be treasured, but, in practice, technology transfer often looks less like a treasure hunt than like a game of catch—with a time bomb.

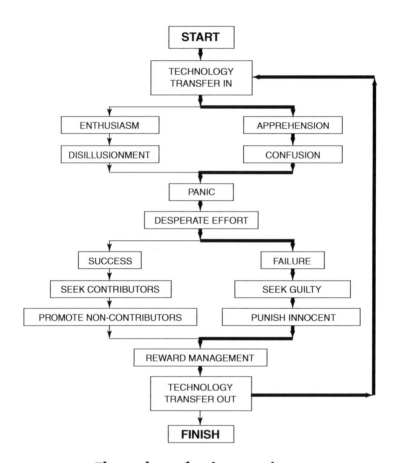

Flow chart for innovation.

The nature of that time bomb is partially revealed by the problems encountered in the Environmental Engineering Group at Ultima. Even greater insight might be provided by a study of events after the project was transferred to the Chemical Engineering Department.

Such case studies have provided a unified picture of the innovative process as viewed from within a small group in any large innovative organization. This is summarized in the simplified flow chart for innovation on the facing page.

Innovation starts with a concept (technology) that is transferred to a group for preliminary development work. The boxes in the flow chart define the various stages of activity, and the arrows indicate the transitions from one activity to another. Heavy arrows represent the paths most frequently followed.

We are not told if Ralph Striker initiated his pollution control project with *enthusiasm* or *apprehension.* But he did *panic* when Proofsworthy suggested that an entirely new approach would be needed. He put all the project personnel on overtime in a *desperate effort* to achieve success, but the result was *failure.*

George Clearwater set up an early morning meeting with the guilty project leader and planned to punish him. Ultimately, however, he chose to *punish* Proofsworthy and *reward* Striker.

Their plan of action called for transfer of the project to the Chemical Engineering Department. If all went as well as Clearwater and Striker hoped, that department would begin work on the pollution control project with *enthusiasm.*

However, if Proofsworthy or someone else were to reveal the true problems, the project would be received with considerable *apprehension.* In either case, work in the Chemical Engineering Department would progress through the various stages in the flow chart until it was time to *reward* management and transfer the project *out,* to another group.

When *technology-transfer-out* is not followed by *technology-transfer-in,* the innovative process is complete. This implies that the original concept is now in use—or that the proposed innovation is *truly finished.*

Exuberance

The importance of choosing good career strategies has been emphasized in previous chapters. It is equally important, however, to project the correct attitude throughout one's career and to understand the role of exuberance in technical and financial endeavors.

The financial markets were jolted in December 1996, when Alan Greenspan suggested that "irrational exuberance" was driving stock prices above rational levels. As the highly respected chairman of the U.S. Federal Reserve Board, Greenspan knew his every word would be scrutinized.

Stock prices dropped precipitously in the first few hours of trading the next day. But by the end of the day, stock prices had mostly recovered. During the next three years, exceptional market gains were recorded. The widely reported Standard and Poor's 500 Index, for example, rose by 34, 29, and 21% in those three years. Only then did a significant and prolonged drop in stock prices occur.

Was Alan Greenspan right, but his timing wrong? Or because he carefully avoided making a specific timing prediction, was he completely right? And what does irrational exuberance in the stock market have to do with high technology and other topics of this book? The answer to the last question is, "Plenty." High-technology companies—especially those involved in building or benefiting from the World Wide Web—were major factors in creating the stock market bubble that finally burst in March 2000.

MAKING MONEY

In 1990, the World Wide Web was little more than the vision of one man, Tim Berners-Lee. Six years later, the possibilities the Web offered for sharing information worldwide and for supporting entirely new forms of business had made it a primary driver of Internet growth and a major focus of high-technology investors.

Recognizing this rapidly changing environment in November 1994, Bill Gates announced that the Microsoft Corporation would enter the online service business and that supporting software would be part of Windows 95. Earlier that year, the Netscape company was formed by Marc Andreesen and others to improve and commercialize the popular Web browser software they had created while employed by the National Center for Supercomputing Applications (NCSA) at the University of Illinois, Urbana.

To obtain the financial resources needed to compete with Microsoft and other large companies, Netscape made an initial public stock offering (IPO) in August 1995. Netscape had never made a profit, but its prospects were good. The offering price was $18 per share. By the end of the day, the price had been bid up to $71 per share. The resulting market value for Netscape was over $4 billion.

By the time Greenspan spoke of irrational exuberance, the stock market was loaded with new or recently redirected information-technology companies. Many of the newer companies were overpriced because there was no way to evaluate them. Indeed, some entrepreneurs with unproven technologies and flawed business plans discovered that an IPO provided the ultimate form of technology transfer.

Even many long-established companies were overpriced based on any standard measure. Consider, for example, the case of Cisco Systems, which had been making routers for the Internet since the mid-1980s. When the market peaked in 2000, Cisco Systems was enjoying a very profitable year. Even so, its stock was selling for 150 times its annual earnings. It was a clear case of irrational exuberance and provided a money-making opportunity for anyone prepared to grasp it.

When IPOs are announced, outsiders focus on the price of the offering and the money to be made during the first few days of trading. Insiders concern themselves with how much of the pie has been reserved for them. Following the market downturn in 2000, IPOs became less common, causing small technology companies increasingly to turn to larger companies to buy them. This practice is quite common because many large companies have learned a basic Law of Invention:

It is more expensive to create inventions
than to buy small companies that have them.

Whether a small company obtains money from selling out to a larger one or through an IPO, insiders are primarily concerned about getting their "fair share." Useless advice is widely available on this subject. The most valuable advice is expressed quite simply by the Rule of the Deal:

When money deals are made,
get a seat at the table.

PUSHING TECHNOLOGY

Irrational exuberance is best known for driving financial markets up, but it is also important in pushing technology forward. How else can one explain the large number of technologists willing to sign on to advanced technology projects, despite knowing that the majority of such projects fail? People sign on, not just for money, but for the sheer pleasure of solving problems never solved before and because irrational exuberance causes them to believe that their contributions will make the big difference.

A thoughtful individual, when asked to lead a project that seems certain to fail, may wonder if it is wise to accept. The correct response is almost always, "Yes." There is, after all, some chance of success. Furthermore, there is a stigma attached to saying, "No," because, as stated in the Law of Leadership Potential:

> Those who see problems become whistleblowers;
> those who see opportunities become leaders.

Project leaders should always monitor the group's progress and obey a primary Rule of Project Leadership:

> When leading a high-tech project,
> keep your eyes on the exit doors.

Leaders of hopelessly doomed projects should also read again the chapter on "The Law of Failure and Putt's Ploy." Failure of a project should never become a failure for the project leader.

METHOD OF RATIONAL EXUBERANCE

Irrational exuberance has been shown to help push technology forward. It was most likely a good elixir for technical giants of the Information Age such as John von Neumann, Bill Gates, Marc Andreesen, and Tim Berners-Lee. But it is a bad elixir for the vast majority of technical people who have more limited capabilities. The elixir of irrational exuberance will only cause them to chase one mirage after another.

For those who recognize their own limitations, this book offers a proven road to winning that I call the Method of Rational Exuberance:

> With each decision you make,
> publicly exude enthusiasm and optimism;
> privately calculate your chances for personal success.

A successful career is practically guaranteed to anyone who combines the Method of Rational Exuberance with knowledge of the laws, corollaries, rules, and tricks of the trade revealed in this book.

Part Two

THE SUCCESSFUL TECHNOCRAT

Innovation

A successful technocrat must be known as a person of great knowledge and as an innovative leader. Isaac M. Sharp's years in college, in graduate school, and as a university professor were all directed toward this goal. But his first opportunity to test his methods and to acquire a reputation as an innovator did not occur until he was hired by the president of the Ajax Vacuum Pump Corporation to manage its research and development activities.

On Dr. Sharp's first day in this new assignment, he chatted briefly with the president about the importance of innovation. Then he moved into his new office and began outlining a strategy for his first conference with two key subordinates. He realized he should not get involved in technical details of vacuum pumps, about which Jones and Smith would know far more than he. Sharp planned to take early command of the discussion and to assert his authority. Yet he had to avoid antagonizing the men who had been the top two technical managers before his arrival. He would need their support.

Now and then, as he reviewed his strategy, he looked up from his desk to admire his office. The thick carpeting, mahogany desk, leather chairs, large sofa, and coordinated draperies were elegant but not ostentatious. The newly painted sign on the door read simply: I. M. Sharp, Vice President for Research and Development. It was quite a step from the university.

Next door, in an office of similar size, sat Allen Jax, president and founder of the Ajax Vacuum Pump Corporation. As Mr. Jax looked over the monthly status reports, he thought back to those years of rapid corporate growth that followed his introduction of the novel Ajax rotary pump mechanism. That single innovative concept, more than anything else, had caused Ajax to become a leading supplier of large industrial vacuum pumps.

When sales began leveling off some years ago, he added several people to the engineering staff in order to get other innovative ideas. He had especially high hopes for Dr. Smith, a bright young man who had completed his Ph.D. thesis on vacuum systems at the local university. Mr. Jones, the chief engineer, was quite satisfied with Smith, who had already made significant contributions to numerous product improvements.

President Jax, however, was not satisfied. He wanted the Engineering Department to do more than make minor improvements to the product line. Although these improvements did keep Ajax's pumps competitive, they were not dramatic innovations that might catapult Ajax well ahead of its competition.

It is not surprising, then, that Allen Jax had become enamored with a brash young man named Isaac M. Sharp, whom he had met during negotiations for the supply of vacuum pumps to the NASA space program. Dr. Sharp was introduced to him as a world-renowned expert on vacuum systems and space research. He was a professor of physics at a leading university, did consulting for several companies involved in the space program, and served on numerous government advisory committees.

On this day—which was Dr. Sharp's first day at work—Allen Jax was already experiencing a sense of excitement regarding future innovations. Hiring Dr. Sharp had been a major coup for Ajax.

THE SEARCH FOR IDEAS

That afternoon, when Jones and Smith arrived in Sharp's office, they were awed by the surroundings and happy to let the new vice president do most of the talking. Sharp immediately launched into a discussion of innovative research. He explained the importance of placing greater emphasis on dramatic innovations rather than on making minor improvements to the present line of vacuum pumps.

He cited several innovative concepts in the NASA program, to which he said he had personally contributed. He further suggested that Ajax should somehow profit from the vast government expenditures. Then he discussed activities of some companies he had consulted for that had initiated innovative outer-space projects.

"Not too many years from now, private companies will have their own space stations and rocket ships to ferry employees back and forth between space stations and Earth," Sharp asserted. "The opportunities for private enterprise in outer space are unlimited." As a first step for Ajax, he asked Jones and Smith to consider ways to make commercial use of the vacuum in outer space.

Dr. Smith at once thought of the possibility of fabricating integrated electronic circuits in a spaceship orbiting the earth. Such fabrication was conventionally done in expensive vacuum chambers from which air and other contaminants were removed by Ajax's vacuum pumps. To be economical, the money saved by using the free vacuum in outer space must be greater than the money spent in shipping parts between earth and the space factory. These savings would also have to be large enough to pay for the high cost of constructing and maintaining a space factory.

That evening, after dinner, Smith did a preliminary analysis of the concept with some scribbled calculations on the back of an old envelope. He was a bit surprised and greatly elated that his calculations suggested the concept might be attractive. He could not resist mentioning this result to the vice president the following day, especially since it would help divert attention from his early afternoon departures for the golf course with Mr. Jones. He further stated that he had discussed his ideas with Jones, who, he said, thought they were quite interesting.

When he actually did mention the idea to Jones at a subsequent golf game, the chief engineer expressed surprise that the large cost of assembling and operating a manufacturing facility in outer space could be offset by the availability of the free vacuum. He noted the billions of dollars that had been spent just to assemble the modest-sized international space station.

Jones did not have time to think about the idea again before he and Smith were asked to discuss the project with Dr. Sharp the following week. Word had begun to spread among top management of the plan to put Ajax into space research. The vice president was excited and eager to keep informed.

A PROJECT IS PROPOSED

As Dr. Smith explained his ideas, the vice president was clearly skeptical, and Jones became increasingly uneasy. He pulled out his golf card and began some hurried calculations on the back. Smith's calculations appeared to have errors of several factors of ten. If this were true, the venture would be foolish. He was tempted to point this out during the discussion but hesitated because he and Smith should have completed this simple analysis long before presenting it to the vice president. Furthermore, he had given the subject very little thought. It was possible that he had overlooked something contained in Smith's more detailed analysis.

When asked to give his views on the subject, Jones simply stated that more detailed analyses now in progress suggested the costs might be less favorable than Smith had indicated. With some quick, on-the-feet thinking, he further noted that another approach was to bottle the vacuum in space and bring it back to earth for use in processing electronic components.

Jones's statements removed much of the vice president's original skepticism. After all, once a project had progressed to more detailed evaluations and considerations of alternative approaches, it could not be dismissed lightly. He, therefore, instructed the two to begin charging their time to the new Space Vacuum Project at the end of the week when their present project was to be completed.

Unknown to the vice president, however, they had to use the first five months of the time allotted to the Space Vacuum Project to complete their previous project. This left only four weeks to prepare the first semiannual report on the new project.

Within one week of actual work they were convinced, by elementary analyses, that space flight was much too expensive to justify its use for fabricating electronic circuitry in space or for bringing vacuum back to earth. They spent the next three weeks in near panic looking for alternative approaches or ways to justify the six months supposedly spent to understand such elementary considerations, but to no avail.

DISASTER IS AVERTED

Only the inventiveness of the vice president saved them from immediate disaster. After briefly listening to a preliminary version of their report, he asked if they had considered compressing the vacuum in space before shipping it back to earth. In this way, he reasoned, smaller containers could be used to bring back the same amount of vacuum more economically. In support of his idea, he noted that natural gas was compressed into cylinders to reduce the cost of shipping it. Even garbage was often compressed before being hauled away.

His idea, he asserted with evident satisfaction, was the type of innovative thinking that was so important to Ajax's future.

Jones and Smith were shaken by the vice president's enthusiasm for this idea. It revealed, among other things, Dr. Sharp's complete ignorance of the nature of vacuum. He apparently did not know that vacuum consists of space from which almost all molecules of air and other gases have been removed. To compress it would only place the few remaining molecules into a smaller and smaller space until their concentration would be as high as on earth. At this point, there would be no vacuum at all, and the contents of the container would be worthless!

Rather than correcting his error during the meeting, Jones and Smith prudently requested time to consider the vice president's suggestion in greater depth.

Success

That night, with the aid of several martinis, Jones and Smith contemplated their options.

They could explain to the vice president that compressing vacuum made no sense and that there was no economic way for Ajax to utilize the vacuum in outer space. With this straightforward evaluation, the Space Vacuum Project would terminate with six months' effort charged to it—a relatively minor failure when measured against Ajax's total research and development effort. The blemish on their records would be somewhat more lasting, however, because they would have to confront the vice president with the foolishness of his idea.

An alternative approach would be to recommend increasing the Space Vacuum Project from two to twenty persons for the next year and to make compressed vacuum the number-one approach. Although doomed to failure, and at a much greater cost, this second approach had much to recommend it. First, it postponed the day of reckoning by one year. Second, it would place Jones and Smith in charge of a major project. Third, and perhaps most important, by incorporating Sharp's idea, the vice president would become personally committed to their program. Such a commitment would be particularly important in a project for which technical failure could be predicted from the beginning.

Putt's Law and the Successful Technocrat. By Archibald Putt
Copyright © 2006 the Institute of Electrical and Electronics Engineers

The correct alternative was embraced, and Dr. Sharp was delighted the next day to learn that his proposal would become a major part of Ajax's first truly innovative research effort. Smith was placed in charge of theory and analysis, and Jones was given overall responsibility for implementing the new project.

The project gradually built up in size until the requested twenty-person effort was in place. Frequent progress reports were made to Allen Jax, and Dr. Sharp was always present to point out the magnitude of the technical problems already overcome while at the same time cautioning him on the problems ahead. The vice president exuded confidence and noted that success would be achieved sooner if more people with critical skills were added to the project. Thus, by the end of the first year, the twenty-person level had been exceeded, and the project was still growing.

Technical details concerning the project were closely guarded. But the existence of the Space Vacuum Project was well known to members of management and to many senior employees of Ajax. Several employees, who had known Allen Jax since the early years, found opportunities to congratulate him on leading the company into the space age.

Even the leaders of the community seemed to know that something exciting was underway at the Ajax Vacuum Pump Company. Allen Jax increasingly found himself the center of attention at civic gatherings. Intellectuals and professional people were interested in his ideas on science and politics.

Most gratifying to Mr. Jax, however, was a meeting arranged by several of the major stockholders. They expressed their desire to be given the first opportunity to buy new stock in Ajax if the company should require additional funds to carry out its aggressive technical plans.

REWARDS OF INNOVATION

On the other hand, Mr. Jax was becoming increasingly aware of his own limitations, especially in the rapidly moving fields of technology. He found much of Dr. Sharp's discussions to be over his head. Even the briefings on the Space Vacuum Project by Jones and Smith were frequently beyond his limited training.

He was also becoming aware that Sharp had a remarkable grasp of business. Thus by the second year of the Space Vacuum Project, Allen Jax chose to retire from the presidency and confidently turned the reins over to Isaac M. Sharp.

This announcement came as no surprise to those in the know, for Dr. Sharp had done much to stimulate innovative research for Ajax's future growth. It was also no surprise when Sharp's previous position of vice president for Research and Development was split into two positions, with Mr. Jones becoming vice president for Development and Dr. Smith promoted to vice president for Research.

The official announcement further noted that such excellent progress had been made on the Space Vacuum Project under the direction of Sharp, Jones, and Smith that it would now be treated as a development program rather than as a research project. Thus, it came under the jurisdiction of Mr. Jones, the new vice president for Development.

The only real surprise came when Jones subsequently announced his selection of the person to be responsible for completing this all-important development effort. It was Bob Plodder, a practical sort of engineer, whom many had thought to be one of the least-favored subordinates of Jones or Smith.

A Lucky Engineer

Bob Plodder went home that evening elated with his new responsibility. He and his wife celebrated by opening an expensive bottle of wine for dinner.

It was not the relaxed and festive occasion they had both anticipated, however, because Bob could not stop thinking about the many technical problems associated with the Space Vacuum Project. All during the evening and long after they went to bed, Bob continued to mull over his new assignment.

In the past, he had been given only routine engineering chores for the project. Now he would have to understand all major technical problems. It was his responsibility to assure that a practical and profitable way was developed to use the vacuum in outer space. A brief discussion with Jones that day had already revealed several problems of considerable concern. While he could see ways around these problems, he was much less confident that the result would be economically attractive.

As he thought over the project that night, his plan of action developed. He would make a list of all the technical problems, as well as the most likely solutions. Then he would determine the time and staffing required to solve the problems and present these findings to management. This presentation would help him establish the technical direction of the project and get the necessary resources.

With his plan of action defined, Bob finally fell asleep. In what seemed like minutes, the alarm went off. He forced himself out of bed, ate a hurried breakfast, and headed off to work—tired and sleepy, but eager to begin his new assignment as program development manager at Ajax.

Within a week, Bob and the engineers working for him had identified a long list of technical problems, but their list of solutions grew more slowly. In a number of areas, there appeared to be no solutions at all without major technical breakthroughs. Such breakthroughs should not be expected with his limited resources, and could not be guaranteed even with a much larger effort.

It was clear to Bob that the engineering work up to now had been confined to easy problems. The really hard problems were ignored, unidentified, or postponed. This practice could no longer be continued if a practical space factory were to be developed.

A REPORT TO MANAGEMENT

By the end of the month, Bob Plodder was ready with a report to management. It was filled with facts and analyses that would reveal for the first time just how long and expensive the remaining development effort would be. It was not a report he looked forward to giving, for bad news is seldom well received. But he was pleased with the thoroughness of his efforts and the technical competence it demonstrated.

Mr. Jones and Dr. Smith showed up promptly for Plodder's presentation and apologized for Dr. Sharp who, they said, would be arriving later from Washington where he was attending a technical advisory committee meeting. They suggested that Plodder begin his presentation and they would give Sharp a quick briefing when he arrived.

Bob began by listing all the technical problems. It was a longer list than Jones or Smith had anticipated. However, considering the manner in which the project had been initiated, they were not altogether surprised. What did surprise them was the large number of potential solutions that Plodder and the other engineers had been able to develop. Some of the more ingenious solutions so intrigued them that they were privately more optimistic about the final outcome than they had ever been before. They enthusiastically joined Plodder in a lengthy discussion of these ideas, as well as of the many problems for which no solutions had been found.

Dr. Sharp did not return from Washington during the meeting, so Jones and Smith made an appointment with him the next day. Jones began the meeting by listing the innovative solutions proposed by Plodder. The ingenuity of some of these ideas had sufficiently impressed him that he regretted not being able to take more personal credit for them. But he did take advantage of the opportunity to note that the excellent progress made by the Space Vacuum Project was indicative of the effective engineering effort he had structured following the recent reorganization. The separation of Research from Development had permitted him, as vice president for Development, to emphasize the need for fast response to engineering problems.

PLODDER LOSES HIS SUPPORT

Sharp, however, showed little interest until Jones got into the area of unsolved problems. Then he responded with anger.

"There is only one thing you've convinced me of," he said to Jones, "and that is you've selected the wrong man for this job. The three of us did the innovative work. Then I established the Space Vacuum Project, obtained funding for it, and made it the number-one project at Ajax. It took real courage to start an aggressive blue-sky project like this—but I did it. Now, this engineer, this engineer you've put in charge of the project, tells us he can't solve a few engineering problems without more people, more money, and more time—but I tell you these are not the things he needs more of. What he does need more of is engineering competence, hard work, and a bit of confidence."

"You're absolutely right about that," responded Jones, who quickly reversed his earlier position. "As you may recall, I have been quite concerned about having him in charge of this project from the beginning. He's not a strong technical person or a good manager, but I had little choice. Our more experienced project leaders were needed for upgrading our vacuum pump designs.

I've tried to work with Plodder to get him on track and to help him find solutions instead of problems—but there is only so much I can do and still keep up with my other responsibilities."

"The Research Department has been doing all it can to help," Smith interjected. "Plodder's group probably wouldn't have found any of the solutions they did without us. Unfortunately, I can't free up any additional resources to support the effort, but I'll try to put more time on it myself. Plodder's problem may be more a lack of technical insight than lack of people to do the work."

"I wish I could roll up my sleeves in the laboratory and go to work on this project myself," responded Sharp. "Unfortunately, my administrative responsibilities here at Ajax don't permit me to get into the lab, and, of course, these advisory committees in Washington take a lot of time. I'd like to get off some of them, but really competent technical talent is so scarce that I just couldn't let the country down. So I have no choice but to leave the Space Vacuum Project to others."

"I know just how you feel," responded Jones. "It's damned frustrating to have a second-rate team on an important project like this. With the belt-tightening here at Ajax, there is no chance I can add additional resources. I may even have to remove some of the engineers from the Space Vacuum Project to beef up our vacuum pump design effort. We've got to make our bread-and-butter products more competitive."

"With the financial crunch Ajax is in right now, you've got to give that top priority," Sharp agreed.

"I guess," concluded Jones reluctantly, "I'll have to tell Plodder and his group to get down to some hard engineering work. It's time he quit wasting our time with lists of unsolved problems and requests for more money."

"That's the thing to do, all right," agreed Sharp. Then he added, "You tell Plodder there are lots of engineers who would like to have his job. I expect his next report to concentrate on progress—not on problems. This project represents a great opportunity for Ajax and for all the people working on it. Plodder is damned lucky to be in charge of it—he's a very lucky engineer."

Genesis of a Manager

Nothing in Sharp's childhood suggested he would grow up to become a national figure in science and technology. He had little interest in building model airplanes, using chemistry sets, or playing with electronic devices, and he was only moderately good in mathematics.

His decision to enter the university with a major in science was accepted by his family and high school advisor as one of several reasonable choices. He had demonstrated no unusual talent in any area, yet he had performed acceptably well in all. His family had always expected him to attend college, and because they were of modest means, he would have to support himself after college.

He selected science because it seemed to be more directly applicable to earning a living than subjects such as English, history, or music. Furthermore, he had been advised that it would provide a good background for any field of engineering.

His grades were good during his first year in college, but he had to work hard and was one of the first to ferret out and study old exams. It was not until his sophomore year that he became panicky about his choice of science. One of his required courses included laboratory experiments. Most of his classmates had experience with laboratory equipment because of childhood hobbies. Sharp had not. They looked forward to the laboratory work. He did not.

As his classmates selected partners for the first experiment, Sharp held back and was ultimately assigned to work with two fellows who had already teamed up. They seemed to know how to wire things together and how to take the data, so he sat back and recorded the results in a notebook. He also fetched equipment from the stockroom and read ahead in the manual to see what measurements his partners should take next.

A TASTE OF SUCCESS

The experiment was completed quickly, and the partnership continued throughout the semester. The other two preferred to use the equipment and to make the measurements without interference by Sharp. Moreover, Sharp expedited matters by recording data and reading ahead in the manual. Knowing what needed to be done next, he tended to organize the activity.

By not knowing how to use the equipment, Sharp had in effect become the manager. It was an experience he did not forget. He enjoyed this first taste of *competence inversion*—playing the managerial role while knowing less about the work than the others. While he had never seen a statement of Putt's Law, he was beginning to understand its ramifications and how to benefit from them.

Ph.D.-itis

Isaac M. Sharp began considering the importance of getting a doctoral degree during his junior year in college. His need for a Ph.D. resulted from a number of changes in the technical community after World War II. Prior to that, Ph.D.s were sufficiently rare that they were unnecessary, perhaps a bit frivolous.

During and after World War II, the importance of the technical contributions of such persons as Dr. Einstein, Dr. Oppenheimer, and Dr. Teller not only gave technical leaders much prominence, but also popularized the Ph.D. This popularization was further enhanced by the importance of transistors, which were invented in 1947 by Drs. Bardeen, Brattain, and Shockley.

When Sputnik was launched by the Russians in 1957, educators and administrators, who were responsible for advancing United States technical programs, found they could report their progress best in terms of the number of Ph.D.s they were producing or hiring. Thus, the social status, economic position, and total number of Ph.D.s in the United States increased dramatically during the next few decades when Sharp was making his own educational plans.

One result of these changes was the outbreak of a new disease in scientific laboratories throughout the country. It was known as Ph.D.-itis. Persons afflicted with this disease tended to blame all their failures on their lack of a Ph.D.

Some Ph.D.-itis victims responded by taking jobs in personnel, sales, or manufacturing. Those who remained behind in research and advanced development slipped into a role of increasing subservience to their new masters, the Ph.D.s. Some ambitious individuals tried placing a "doctor" in front of their names without the usual formalities.

A few adopted the well-established gimmick of advertising what they could not correct by implying they were too good to need a Ph.D. Astute practitioners of this gimmick cited people in their field who had done especially well without a Ph.D. Then they clinched the argument by noting that Bill Gates had dropped out of Harvard even before getting his bachelor's degree.

The various ploys, successes, and failures of the scientists and engineers without Ph.D.s during this period are of some historical interest. For Sharp, however, it was sufficient to note that a Ph.D. had become for the scientist and engineer what the union card was to the plumber or electrician. A few might succeed without this "union card of technology," but for the majority, the chances for financial reward and success would be dramatically reduced without it.

In the final analysis, Sharp concluded, there was no good cure for Ph.D.-itis other than a Ph.D. He decided to become inoculated as quickly as possible.

THE INOCULATION

Sharp's approach to selecting a thesis topic and thesis advisor was not that of the typical graduate student. His methods and priorities clearly showed his promise as a manager. He recognized how important the side effects from a Ph.D. inoculation might be, so he examined them carefully. Most informative was his examination of past Nobel Prize winners.

He learned that half of the United States Nobel Prize laureates in physics, beginning in 1901, had worked under a previous winner. Was it just a matter of good selection and training, or did a previous winner's recommendation unduly influence the award? Whatever the reason, it was clear to Sharp that a person desiring to become a Nobel laureate should begin by working with one. His chances would be enhanced one thousand-fold by this single step.

Sharp, however, recognized his limitations. Under the best of circumstances, the chances of winning a Nobel Prize would be small, even for someone who had substantially more intelligence and scientific curiosity than he. Furthermore, winning the Nobel Prize was not the only way to achieve prestige and influence in the technical community.

From his study of the Nobel laureates, Sharp concluded that it would be important for him to study under a person with the right skills and the desired position in society. Thus, while most students first select their area of study and then a thesis advisor, Sharp reversed the procedure, concentrating first on finding the right advisor.

He soon settled on Professor Grossmann, a truly big man at the university. Grossmann was in charge of several government contracts, spent a great deal of time consulting, and attended numerous committee meetings in Washington. He was a professor with little time for the technical details of the work of his graduate students, all of whom were supported by his government contracts. Sharp was particularly pleased that Grossmann would be too busy to come into the laboratory where he might learn how inept Sharp really was.

Professor Grossmann, for his part, was delighted to have one more low-cost graduate student, often referred to as slave labor, to help carry out his research activities. His only fee for supporting Sharp's research would be his position as coauthor on any resultant publications.

Regarding the selection of a thesis topic, Grossmann noted that the surest way to get high-salaried offers of employment from industry was to select a topic of immediate practical interest. Once hired, such a person could expect to become involved quickly in industrial advanced development programs which, if successful, would lead to rapid promotion.

On the other hand, Sharp noted that the failure rate of advanced development programs was very high, making this commonly chosen method for getting a quick start statistically unpromising. By the time a person had been involved in enough projects to have a successful one, he would probably no longer be regarded as a person with a promising future.

Being a man of considerable ambition and caution, Sharp decided to select a thesis topic on an esoteric subject of no practical value. For this choice, he was treated with a great deal of respect. The faculty referred to him as a truly dedicated scientist, and he himself marveled at how profound his pronouncements on this esoteric topic began to sound.

Before making the thesis topic selection final, Professor Grossmann suggested that he and Sharp discuss it with an outstanding member of the technical community. Grossmann ultimately arranged this discussion to be with a past Nobel Prize winner.

This pleased Sharp and confirmed the wisdom of his choice of Grossmann to be his advisor. Reference to this discussion would provide adequate defense for the choice of the topic. Furthermore, when the thesis was finally written, he could acknowledge the Nobel laureate for his "aid in selecting the topic" and for "helpful discussions." This implied close association with a Nobel laureate would greatly enhance his own image.

Jargonese

Isaac Sharp was distressed that a fellow graduate student with a thick foreign accent was regarded more highly than he. How could this be, when the foreign student did no better than he on written exams? Was there some advantage in having a foreign accent?

To answer these questions, Sharp became an observer of verbal communications between faculty members and the many foreign students in science and engineering. He quickly became convinced that faculty members routinely gave foreign students the benefit of the doubt concerning the intended meaning of their statements.

Taking advantage of this phenomenon, some foreign students had learned to exaggerate their accent when they were not sure of the answer. Much to his delight, Sharp even observed that the slight German accent of his thesis advisor became thicker as the subjects became more difficult.

"What tool," he asked himself, "could an American-born scientist develop to meet this foreign challenge?" One could cultivate a foreign accent by spending a year or two abroad or perhaps a few months in a language school. But he dismissed that idea as impractical. How could a person born in Decatur, Illinois, explain a foreign accent?

A SATISFACTORY SOLUTION

The only satisfactory solution was to develop his skill in the language of Jargonese. This language consists of the grammar and vocabulary of any natural language, upon which is superimposed an impressive collection of specialized technical terms. Jargonese may be as old as language itself. It has been in existence ever since the first witch doctors did their incantations to drive away evil spirits. And it persists today in Latin terms in law and medicine.

As physics and chemistry became more precise and their scope broadened, it became necessary to define new terms. Many of these describe concepts beyond the scope of normal human observation. Thus, the discussion of chemistry or physics by experts is likely to contain enough specialized terms to be incomprehensible to the general public. The weakness of these terms as a basis for Jargonese is that they are well defined and well understood by experts.

True Jargonese must be able to be made obscure even to the expert. This occurs naturally only when a subject is evolving so rapidly that new terms are invented daily. It is most effective when the subject is not well structured so that contradictory and ambiguous terms can coexist for months or even years.

Such a subject has been born of the information processing industry. It is called Computer Science and its dialect is Computerese. With help from the Internet, the power, diversity, and utility of Computerese continues to grow. It is acceptable in science and engineering laboratories, at missile launching sites, at cocktail parties, and during most board of directors' meetings. It is clearly the most promising form of Jargonese!

The vocabulary of all forms of Jargonese consists largely of nouns with only a smattering of adjectives and verbs, and it is devoid of any connectives, articles, pronouns, or prepositions. Thus, it is not always possible to construct sentences that sound profound throughout. In fact, avoiding the appearance of rank stupidity may at times be difficult.

This problem can be solved by chewing on a pencil or inserting a pipe or cigarette into the mouth when speaking the more technically questionable or mundane portions of each sentence. Some practitioners of the art build up a melodic rhythm

Computerese in use at the cocktail party.

in which each sentence begins with a ring of earnest profundity and fades into a tantalizingly obscure mumble.

ADVANCED METHODS

Advanced practitioners also make use of a variation of the minimax strategy of Game Theory, in which players strive to minimize their maximum loss. The variation of that strategy can be described simply as shifting the conversation to a field in which you have maximum competence and your listeners have minimum knowledge. The selected subject should also be one in which all participants have, insofar as possible, maximum interest. By employing the minimax strategy, the successful technocrat can maximize his apparent brilliance with a minimum of knowledge.

Finally, the inscrutable smile is the most powerful and dangerous tool employed by advanced Jargoneers. Under some circumstances it may be interpreted as implying agreement with the speaker, whereas in other circumstances it suggests complete disdain. It can be used to suggest the former to one listener, while simultaneously suggesting the latter to another. When asked a direct question about which you know practically nothing, an inscrutable smile may suggest you have too much knowledge of the subject to give an answer simple enough for the listener to understand.

Sharp's observations convinced him that a tool as powerful as the inscrutable smile should be used only by the most skilled practitioners, for it can raise in its victims passions of hate and an overwhelming desire for revenge. Furthermore, if used incorrectly, the unfortunate practitioner can become an object of contempt or ridicule.

Acknowledgments

Sharp's laboratory work was not going well. His advisor was seldom around, and he was spending much of his time practicing Jargonese and other tools of the trade with fellow graduate students. He was thus surprised and concerned when Professor Grossmann suggested he present preliminary experimental results at the next meeting of the technical society.

He almost objected that he had no reliable measurements and could not interpret his results. However, as Grossmann talked on, Sharp realized how anxious his advisor was to attend the conference and that none of the other graduate students were ready. Presenting a paper was necessary if the trip was to be paid for by the government contract. Sharp therefore agreed to get together the next day to see if Grossmann felt his results were adequate.

Professor Grossmann was pleased with Sharp's eagerness. Early publication would help to prove that his research projects were the most productive on campus. When he saw Sharp's results, he was not the least bit surprised at how jumbled they were. He had been confronted with this many times before and prided himself on his ability to take inconclusive data and develop a presentation that made them appear significant. A technical conference was an ideal place to practice this art because the results would be published only as a brief abstract in the *Conference Proceedings*.

Putt's Law and the Successful Technocrat. By Archibald Putt
Copyright © 2006 the Institute of Electrical and Electronics Engineers

The two spent several hours huddled together viewing the data, first one way and then another, until a plausible explanation and theory were selected. Then Grossmann arranged for a rehearsal of the talk before two of the more respected and conscientious members of the faculty. Meanwhile, Sharp spent his first really solid week in the laboratory, trying to get additional data to support the proposed theory.

Sharp stayed up late the night before the rehearsal to organize the new and old data in graphs and tables to discuss with Professor Grossmann in the morning. His advisor was pleased with all the new data. After making several suggestions for improving the talk, Grossmann accompanied Sharp down the hall to a lecture room where Professors Good and Stockton were already waiting.

A FUNDAMENTAL PROBLEM

Following the talk, which took fifteen minutes instead of the allotted ten, Grossmann began suggesting ways to eliminate the extra five minutes. The other two professors were troubled by a more fundamental problem. They pointed out that the proposed theory was in conflict with the data in a number of respects. This had not been noticed by Sharp or Grossmann because it was evident only from Sharp's more recent experimental results.

Grossmann proposed they discard these new results on the grounds the data may have been taken too hurriedly. Sharp did not object, although further discussion revealed that he believed the more recent results were his most reliable ones. He and Grossman continued their discussion without resolving the issues.

Finally, Professor Good hesitantly interrupted. He said he had recently received a preprint from a British experimentalist that contained measurements consistent with Sharp's. Moreover, the preprint referred to a new theory that might help explain the data. Professor Good had not yet studied the theory, but he planned to do so and would be glad to discuss it with Sharp in the next few days. Sharp and Grossmann immediately accepted the offer. It appeared to be the only hope of salvaging the talk and perhaps even the thesis.

Professor Grossmann was out of town most of the week, so Sharp and Professor Good got together alone. They studied Sharp's data and that of the British experimentalist until they had both results well in mind. Then Good discussed the new theory in such a simple and elegant fashion that Sharp felt he understood it, even though the mathematics would have taken him days to comprehend. He was amazed at the professor's ability to derive so much meaning out of such a complex mathematical paper.

Then they reviewed the data in light of the new theory. They were ecstatic. The data and theory were in perfect agreement!

Professor Good was delighted that the hunch he had voiced during Sharp's dry run had proved to be correct. They worked past dinner that evening and got together several times during the next few days. Never before had Sharp worked so closely with such a brilliant, yet self-effacing person.

HELPFUL SUGGESTIONS AND DISCUSSIONS

Sharp managed to schedule a thirty-minute appointment with Grossmann for a private rehearsal of the much-revised talk just before they left for the conference city. Grossmann was positively delighted with the way the data and theory now went together. Although Sharp did not want to diminish the magnitude of his own contribution, he did feel obliged to indicate how much time Good had spent and how important his help had been. Sharp also asked if Professor Good should be a coauthor on the paper.

"There's no need for that," responded Grossmann. "Professor Good is well known in the field. He has his own graduate students and gets on enough papers so that one more won't make much difference. We'll just put an acknowledgment as a footnote on the paper, thanking him for his helpful suggestions and discussions."

The talk at the conference went well, and Sharp immediately began writing his thesis. This was greatly facilitated by Professor Good's contribution, and Sharp frequently consulted with him, especially since Grossmann was seldom available.

These discussions proved invaluable in preparing Sharp for his formal defense of the thesis before a committee of faculty members.

The final step for Sharp was external publication of the thesis results. Grossmann noted that the thesis contained both experimental results and theory. Thus, it could be divided into two parts, which meant that two papers, instead of one, could be written for external publication.

Sharp's name was listed ahead of Grossmann's on the experimental paper, as is traditional with graduate students and their advisors in this country. However, on the more theoretical paper, Grossmann placed his name before Sharp's. This was appropriate, he reasoned, because the theoretical portion resulted directly from the rehearsal that he had arranged with Professors Good and Stockton.

The very important contribution of Professor Good was acknowledged along with that of Stockton in a note at the end of the theoretical paper, which read simply, "The authors wish to express their appreciation to R. E. Barker, I. M. Good, W. Donnel, J. C. Major, and T. S. Stockton for helpful suggestions and discussions."

Sharp recognized Barker and Major as two members of the faculty who taught Grossmann's classes when he was out of town. W. Donnel was apparently a colleague of Grossmann whom Sharp had never met.

Publish and Perish

Graduate student Sharp, soon to be Dr. Sharp, was justifiably proud at having achieved a conference paper and two formal publications from his first research effort. He was nevertheless envious of his foreign contemporaries who could have twice as many publications with the same effort by publishing each paper in their native language for local consumption and then again in English, with minor modifications, for the international scientific community.

Published articles such as these are used to communicate with other scientists and engineers and to preserve valuable information. The real importance to the scientist, however, is that it is one of the few quantitative measures of performance. The chairman of a department in a university may not be able to judge the quality of work in a specialized field, but he can count the number of papers published.

Even the dean of the school and the president of the university can count. Recognition of this self-evident fact led to the use of the phrase "publish or perish" to describe the requirements for survival in the academic community.

There is tacit agreement among paper counters that one publication per year is the minimum acceptable output, while two to three is quite acceptable. The fact that Einstein published four papers in 1905 makes the number *four* particularly auspicious. This is especially true because one of his papers, on the

Even the president of the university can count.

photon theory of light, resulted in his receiving the Nobel Prize in physics; a second paper created the theory of Brownian motion; a third established the mass-energy equivalence of the now famous $E = MC^2$ equation; and the fourth was his historic pronouncement of the Theory of Special Relativity and the *fourth* dimension of time.

The numbers *seven* and *eleven* are also highly regarded.

Technologists who publish more than eleven papers per year may be accused of publishing merely for the sake of publishing, for it is clear that there is some upper limit to the number of significant contributions a person can make. The record for significant contributions per year may well be held by Lord Kelvin, who maintained an average of fifteen papers per year from 1870 until 1900. Beyond this upper limit, each additional publication will most likely have a negative effect on one's reputation, and it is clear that the point of diminishing returns sets in shortly after the Einsteinian number of four.

The "publish or perish" syndrome in the technical community, combined with human vanity, has caused a printed-word explosion. Looking into the matter further, Sharp learned that since 1750, when there were only ten journals, there had been a regular tenfold increase every 60 years. By 1980, over 100,000 journals were being published. If all the journals printed and distributed that year could have been stacked in one pile, it would have been over 2000 miles high.

OPTING FOR SUCCESS

By opting for success rather than technical contribution, Sharp could take pride in knowing he would not contribute for long to this form of pollution. He knew that a technologist who stakes his fortunes on research and publishing can seldom achieve a position beyond that accorded to anyone else who publishes two to four papers a year.

For greater prestige and financial reward, Sharp chose another path to maximize his rate of rise in technical management. It was a path whose viability was well attested to by the small number of publications, patents, or other evidence of personal technical contribution associated with most individuals in technical management. His chosen path was such that time spent in doing research or writing articles would only serve to reduce his time available for activities essential to a rapid rise in the hierarchy.

Sharp concluded that really successful technocrats do not believe that the "publish or perish" syndrome applies to them. In fact, a wealth of evidence suggests it might better be phrased "publish *and* perish."

Where the Money Is

The most sophisticated bank robber of the twentieth century, Willie Sutton, could boast of robbing one hundred banks and escaping from three "escape-proof" prisons. When asked why he robbed banks, he responded simply, "Because that's where the money is."

The adage of "going where the money is," was one Sharp planned to follow. His goal was to find a position that would lead to the greatest control of research and development funds; for in technology, as in most other fields, the control of money leads most directly to prestige, power, and financial reward.

GRADUATE STUDIES

As a graduate student, Sharp learned that the federal government in Washington, D.C., had provided well over half the money spent in the United States for research and development (R & D) during the Cold War years. But by 1990, the federal government's share had dropped below 40% and was still falling. Industry, where most of the R & D had always been done, was paying an increasing share of the cost. The result was that the country's total expenditures for R & D was holding relatively firm at about 2.5% of the Gross Domestic Product (GDP).

The greater total expenditure by industry did not necessarily make it a better place to start than government, however, because the management of industry was scattered geographically and divided among hundreds of major firms and thousands of minor ones. Only in Washington was the source and control of funds geographically centralized.

Furthermore, the federal government continued to provide most of the money for basic research, much of which was done in universities. Although basic research accounted for less than 20% of the country's R & D expenditures, it was particularly important in Sharp's view. Basic research traditionally led the way to new technologies.

Although the advantages of starting his career in Washington were eminently clear, Sharp noticed some hazards—primarily those of getting lost in the bureaucracy. Therefore, before taking a job in Washington, Sharp felt he should have available several ladders for ascent and at least one hatch for escape.

The most obvious escape hatches were to the industrial firms that received R & D funds from federal agencies. Keeping these hatches lubricated would require that a job in Washington provide contact with key men in these industries and the opportunity to appear essential to the allocation of government R & D funds.

THE IDYLLIC LIFE

The more Sharp weighed the comparative advantages of industrial and government jobs, the greater appeared to be the merit of an interim position in a university. It is a marvelous place from which to study the power structure in the technical community and in which to refine Jargonese and other tools of the trade. Because a university professor has no industrial affiliation, he is a logical choice for membership on government committees that help determine how government R & D funds should be spent in industry.

Moreover, with few schedules to meet, except lecturing to students six to nine hours a week, Sharp would usually be free to attend committee meetings. Membership on important Washington committees is a form of recognition and, therefore, encouraged by the school administration. Grossmann could help put him in touch with the right people.

Committee activity in Washington would not prevent Sharp from consulting for industrial companies. Although the university officially permitted only one day a week for paid consulting, he knew that the more involved he became with Washington committees and professional societies, the more difficult it would be to keep an accurate count.

Thus, he entered into the "idyllic life" of a university professor. With Grossmann's recommendation, he was quickly placed on several Washington committees. Through these committees he got to know important industrial leaders and received several jobs consulting for them. His most lucrative consulting positions were for companies whose businesses were under review by the Washington committees on which he served.

It was also through his committee work that he met Allen Jax, who later hired him to manage Ajax's new Research and Development Department. While at Ajax, he continued his activities on the committees—serving his country and himself as best he could.

Tools of the Trade

The reporter from *Technocracy News* had been waiting in the reception lounge at Ajax for some time before she was finally ushered into Dr. Sharp's office. She was visibly impressed by the large office, plush carpeting, and air of importance surrounding Dr. Sharp. Autographed pictures of senators and cabinet members were prominently displayed along with numerous mementos of his short but distinguished career in science and technology.

As Dr. Sharp finished his telephone call, the reporter apologized for picking such a busy time.

"No problem at all," said Dr. Sharp. "It's like this all the time. It seems they just can't make a decision in Washington without first getting my opinion. Makes it hard to find time to handle the business here at Ajax." With that he sat back and waited for the reporter to state her business.

"As you know," the reporter began, "Congress is holding hearings again on the country's energy policies. The objective of achieving energy independence appears very difficult because the United States consumes 25% of the world's supply of petroleum and has only 3% of the reserves. Of greater concern to some is that, at the projected rate of use, worldwide oil reserves will be exhausted in only 30 to 60 years.

"I understand you were a key technical witness some years ago when these issues were discussed and that you took a strong position on the role of nuclear energy. I wonder if you would clarify your views at that time and tell me what direction the country should take now."

"Certainly," responded Dr. Sharp, "but it's a rather complex issue, you know. It would be difficult to summarize for you my technical recommendations that took half a day before the congressional committee. Have you read the transcript of that testimony?"

"Yes," replied the reporter with obvious satisfaction.

"Fine, then why don't you indicate what parts of the testimony you would like to have clarified?"

"Well, as I recall, two major issues at that time were the difficulty of providing adequate, long-term storage for exhausted nuclear fuel cells and possible accidents at the power plants. The latter concern has been heightened by the threat of terrorist attacks."

"That is correct," Dr. Sharp replied, as he adroitly passed the discussion back to the reporter.

The reporter squirmed, realizing her questions were much too general. She mentally chastised herself for having taken only the minimum required courses in math and science in college, as she desperately groped for more appropriate questions.

Just then, the phone rang. Dr. Sharp picked it up and became involved in a discussion of corporate policy.

When Sharp hung up, the representative of *Technocracy News* made a fresh start. "As I recall it, some experts recommended substantial changes in the design of. . . ."

The phone rang again and Dr. Sharp became involved in another conversation. When the reporter started to sit down to await the call's conclusion, Dr. Sharp motioned her to the whiteboard, which was kept behind doors on the paneled wall. Then he cupped his hand over the telephone mouthpiece and said, "Just go right on talking."

The reporter looked at him incredulously, but seeing he was serious, began to talk again. She tried to sketch on the board some of the graphs contained in the testimony and to recall the various views presented. Meanwhile, Dr. Sharp stared from time to time at her or the whiteboard while he carried on his conversation over the telephone. Occasionally, he muttered a knowing "uh-huh" either to the phone or the reporter. Several times when the reporter felt most uncertain of her statements, her discomfiture was heightened by a quizzical expression or a raised eyebrow of Dr. Sharp.

SHARP TAKES CHARGE

When the phone call ended, Dr. Sharp immediately took charge of the conversation, much to the reporter's relief. He picked up a number of extraneous issues alluded to by the reporter and expounded on them at length. As the words flowed, these extraneous issues took on major importance.

The complexity of the subject made it hard for the reporter to follow, and this was further compounded when Sharp continued talking while lighting his pipe or taking several puffs in succession. She thought Sharp had touched on the redesign issue several times but she was not sure. She hesitated to break in with questions, especially since she could not keep up with the rapidly flowing thoughts.

When, finally, the reporter had the temerity to ask a question, Sharp merely smiled broadly, leaned back in his chair, and puffed on his pipe. After what seemed like a very long time, Sharp broke the silence: "Perhaps you should reread the original transcript and then look over some of the more technical articles referenced in that testimony."

Just then the phone rang again and the secretary's buzzer sounded. Dr. Sharp put the phone to his ear. His brow furrowed and his face took on a most serious expression. He cupped his hand over the mouthpiece and spoke to the reporter, "Sorry to cut you off, but this is the White House on the line so things will be a bit involved for the rest of the day.

"It's a real crime the way Washington has bollixed up the country's energy policy. They are asking questions now that could have been answered easily if they had only spent money on some of the research projects I recommended. They never think ahead. Sometimes I think the only way I can save this country is to give up the presidency of Ajax and go down to Washington to straighten out this R & D mess."

He waved his hand, signaling the end of the interview, and then added, "Regarding the things I've said, you can print anything you want so long as you don't give your source or quote me—that is, if you ever want another interview."

Muddled and confused by the barrage of technical terms, the reporter quickly left the room and boarded the express elevator to the ground floor. She had managed to arrange a crucial interview with a leading member of the scientific community, only to muff it, she believed, due to the inadequacies of her technical background.

She burst out of the building into the cool, refreshing winter air—another unsuspecting victim of Dr. Sharp's tools of the trade.

Thank You, Dr. Sharp

S everal months after the interview with Dr. Sharp at Ajax, the same reporter from *Technocracy News* was admitted to the inner office of one of the most recent recruits to the Washington scene. She greeted the man behind the desk with a cheery, "Good morning, Dr. Sharp. Welcome to Washington!"

"Thank you," responded Sharp. "It's good to see you again. I was sorry to have to cut you off last time, but there were so many problems requiring my attention. I'm sure you understand."

"Yes, sir! It was clear that you were really involved that day. I like to think I may have been in your office at Ajax on the very day the President asked you to take this job."

"Well, that's not a bad conjecture," responded Sharp, without directly confirming it.

"Since I saw you the last time," continued the reporter, "I have become better informed about energy policy issues. So, I have a number of specific questions and won't waste your time with generalities. For example, in reading over your testimony, it was difficult for me to tell whether you were for or against expanding the use of nuclear energy. Can you tell me what your position was?"

Dr. Sharp leaned back in his large desk chair and looked out the window as he puffed on his pipe. "That's a very fascinating subject; and considering the evasive testimony offered by some of the witnesses, I have always been proud of the straight-forward answers I gave at the hearings. Unfortunately, because of the sensitive nature of my new position here in Washington, it would be inappropriate for me to reopen the subject now. I think the important thing is to take a fresh look at all issues and then make the best decisions based on the present state of technology. It would, therefore, be best in this interview if you confine your questions to subjects that do not relate to my past involvement as a government consultant."

"Certainly, sir, I understand," replied the reporter. "Perhaps you would be willing to comment on rumors that Ajax is undergoing severe financial difficulties due to very large expenditures on space research."

"Not a bit of truth to it," Sharp responded. "Ajax's profit has been somewhat reduced recently due to the slump in orders for its vacuum systems. However, we were doing a lot of belt-tightening when I left, so it's a really trim ship at Ajax now. I'm quite optimistic about its future. Mr. Jones has taken over as president. He's a good, aggressive technical manager and a sound business executive as well.

"The Research and Development departments have been combined into one unit under Dr. Smith and the budget reduced somewhat. He is confident they can upgrade the line of vacuum pumps and recover their traditional leadership in markets where there have been recent competitive inroads."

The reporter was not fully satisfied with Dr. Sharp's answer and rephrased her question. "Could you tell me something about the status of the space vacuum effort at Ajax? I understand it was the major development effort up until a number of technical problems were uncovered six months ago, and that this heavy commitment of the technical resources was the primary cause of the profit slump."

"Those rumors about the impact on Ajax's profits are highly exaggerated," responded Sharp with a bit of annoyance. "We did, however, have a number of difficulties with the Space Vacuum Project, which were aggravated by the need to spend more research and development effort on upgrading the product line. But as I said before, we took the necessary steps before I left to assure future profitability."

After a pause, the reporter from *Technocracy News* persisted with the subject. "Is it true," she asked, "that a large number of engineers were laid off as a result of the changed emphasis on that project? One engineer, who claimed to have been in charge of the project before he was laid off, made some rather bitter statements to the press a few weeks ago. I assume you saw that. I believe his name was Plodder, ah, Mr. Robert Plodder."

A TOUGH MANAGEMENT DECISION

"Yes, I did see something to that effect," responded Sharp with reluctance. "Bob Plodder is the man who took over the project when Jones and Smith were made vice presidents. He just didn't have what it takes to manage an innovative, high-risk effort. He never really pursued the project aggressively. He seemed more interested in finding problems than in solving them.

"This created a serious morale problem, so we relieved him of responsibility for the project and Mr. Jones ran it himself for a while. Plodder, however, continued to criticize the technical direction and even complained about the way Jones was managing it. So there really wasn't any choice but to terminate his employment.

"It's the type of tough management decision one has to learn to make. Getting rid of the deadwood is never easy. There were, in fact, several other engineers on the project that we let go during the belt-tightening operation."

"When a situation like that occurs, I guess it's as tough on management as it is on the employees," responded the reporter. "It certainly points out how important it is to have good, aggressive management in an advanced development effort." Then, after further thought, she asked, "Can you tell me how large the project is now and how much importance Ajax attaches to it?"

"I believe the project was cut back to a two-to-three-person effort shortly after I left for Washington," answered Sharp. "Regarding the importance of the project, one should take a long-range view of things. You can be sure that whatever is decided about the project now, Ajax will profit in the long run from the expertise developed and from the various types of technical fallout."

Getting rid of the deadwood is never easy.

"I appreciate your candor regarding that important R & D project at Ajax," responded the reporter. "I'll try to put the whole thing in perspective in my article. I'm sure our readers would also like to hear why you accepted this job and what you hope to accomplish."

WHEN DUTY CALLS

"Delighted to answer that," responded Sharp. "It has been quite a sacrifice for me financially, and, of course, moving my family to Washington has been very disruptive. My wife and I are concerned that all the publicity will be harmful to the children.

"Yet, I felt it was my duty to come. The government really has problems, and someone has to make a sacrifice and get involved if the situation is to be straightened out. When the *president* asks, how can any good citizen refuse? I just hope I will be able to bring my years of experience to bear in an effective way. I wouldn't want my successor to find the same sort of mess that exists today." With that, Sharp waved his hand in a gesture suggesting the interview was over.

"Dr. Sharp, I certainly appreciate the time you have given me," said the reporter as she rose from her chair.

"No problem," responded Sharp. "I always enjoy exchanging views with a perceptive and informed member of the press."

"You've been most helpful," said the reporter with evident appreciation for Sharp's closing remarks. "This country is fortunate to be able to find dedicated people like you to serve in Washington. I'm sure I reflect the view of our readers and of all informed citizens when I say, 'Thank you, Dr. Sharp!'"

Part Three

BASIC PUTT

Laws of Innovation Management

"$\mathbf{H}$ere, you see, it takes all the running you can do to keep in the same place. If you want to get somewhere else, you must run at least twice as fast as that!" So said the White Queen to Alice in Lewis Carroll's classic book, *Through the Looking Glass,* published in 1871. Although Carroll was writing about an imaginary world in the past, his description describes the plight of today's managers of the innovative process.

MANAGEMENT BY OBJECTIVES

Managers are often told they should learn to define their objectives and then manage their organization so as to achieve these objectives. This is known as "management by objectives." It is one of the most popular concepts of management, even though most of the time we do not know what the objectives should be. This is particularly true in the world of technology.

The Mobil Oil Company, for example, was involved for years in research on petroleum and petroleum products. One result of this research was the development of artificial yeast with characteristics superior to natural yeast for many purposes. This placed Mobil Oil in the food processing business, an event that clearly did not result from any predefined objectives.

Scotch Tape was invented for mending books. Its large-scale use for other purposes was never anticipated. And who can recall the original purpose for Silly Putty?

When IBM introduced the 701 computer in 1952, it was thought to be so powerful that perhaps ten and definitely no more than fifty organizations would be able to make use of it. Less than fifteen years later, the IBM System/360 computers, which were up to one hundred times more powerful than the 701, were being installed at a rate of one thousand a month. As the twenty-first century began, personal computers (one hundred times more powerful than large computers of the 1960s) were being sold at a rate of more than 10 million per month.

What objectives should have been set for the management of IBM back in 1952? And what would have happened if these objectives had been followed?

In 1877, George Eastman, a young bank clerk in Rochester, New York, began to plan a vacation in the Caribbean. A friend suggested he take along a photographic outfit, which he discovered, to his dismay, was really a cartload of equipment that included a light-tight tent and other cumbersome objects. Field photography, in those days, required a person who was part chemist and part contortionist. With "wet" plates, there was chemical preparation immediately before exposure and development immediately after—no matter where one might be.

Eastman gave up his trip and began studying photography. Working at night in his mother's kitchen, he began experimenting with dry plates and developed a process that he patented in 1879. In 1880, Eastman began to manufacture dry plates commercially. This venture led to the founding in 1892 of the Eastman Kodak Company, whose products popularized photography and created a major industry.

If George Eastman had been managed by objectives, we can only assume he would have gone on the planned vacation to the Caribbean and taken many photographs, which long since would have faded from memory.

From these cases and many others, the First Law of Innovation Management emerges:

Management by objectives
is no better than the objectives.

Management by objectives would have been easy for Thomas A. Edison. He understood his own objectives, and he never had to work within an innovative bureaucracy. This is just as well, for there is good reason to believe he would not have been able to accomplish nearly so much.

Yet Edison, more than any other person, is responsible for the creation of large organizations whose sole function is research and development. The first laboratory ever devoted to industrial research was established by Edison in Menlo Park, New Jersey, in 1876. There he invented and innovated with the help of many associates. By the time of his death, in 1931, incandescent lamps, phonographs, fluoroscopes, and many of his other inventions were in use throughout the world.

THE INNOVATIVE MOB

The successes of Edison's laboratory, as well as those of the large organizations that developed nuclear energy and transistor technology, have led to the assembling of other large innovative organizations throughout the private and public sectors.

Thus has been born the *technical hierarchy* of innovative technical people—a product of the innovative successes of the early 1900s. So creative and competent in its parts, yet so unmanageable in its entirety, the modern *technical hierarchy* has become a major innovative force.

Within these hierarchies are contained the bubbling creative energies of hundreds to thousands of technically trained people. The inventive and innovative possibilities are staggering, as these individuals vie with each other to be the first with the next discovery. The thrust of this *innovative mob* is enormous, and the job of managing it is virtually impossible. As stated in the Second Law of Innovation Management:

A manager cannot tell
if he is leading an innovative mob
or being chased by it.

**Does a manager lead the innovators,
or is he chased by them?**

One solution to this problem has been found by successful managers who obey the Third Law of Innovation Management:

Stay in the pack until the objectives are clear.

These managers express no strong opinions and make no major decisions until most of the uncertainties are removed. Thus, the hierarchical aging process begins. Organizations that originally were aggressive or innovative begin to move toward a conservative or, worse, to a completely stagnant state.

A SUBTLE FATE

The unofficial objective in a stagnant organization is to stand pat. Anyone bold enough to refuse to stand pat is moved quickly outside the group. A similar fate awaits the nonconformist in a conservative organization. The fate of those who reject management objectives in aggressive or innovative organizations is more subtle.

Alice Highmind frequently refused to accept management objectives. Her reasons were well thought out and clearly stated, so management generally let her pursue her own ideas. Her salary increases were slow in coming, but she was tolerated. After all, her existence in the hierarchy provided proof that individual initiative was appreciated.

On one occasion, Highmind was particularly adamant. It was not just part of the project, but the whole project to which she objected. She stated unconditionally that management's approach would not work.

In order to quiet her, management gave her one assistant to work on her ideas. Then they put the bulk of their resources behind their own approach. Months later, it was evident that the two-person effort was further along than the larger project. Furthermore, problems had been uncovered that supported Alice's original contention.

Management now faced a difficult problem. How could they switch resources to Highmind's approach without admitting to their original error? They began by adding four people to her activity, including a manager for the "new" project. Highmind's assistant was replaced by a more experienced person, and, ultimately, Highmind herself was in serious difficulty for failure to contribute effectively to the new project. The reported inadequacies of Highmind's work were overcome only by a series of aggressive management actions.

Alice was unhappy and puzzled by the outcome. She had expected to be commended for refusing to accept the incorrect objective and for the success of her own initiative. But just the opposite had occurred. She had become an unsuspecting victim of one of the most insidious laws of management, the Law of Insubordination:

> Rejection of management objectives
> is undesirable when you are wrong
> and unforgivable when you are right.

Four Laws of Advice

L earning to give good advice is so important to technologists that special laws of advice have been developed. The first of these is often stated as follows: "The correct advice is the desired advice." However, this form of the law leaves ambiguous whether the recipient wants correct advice or whether the desired advice is by definition the correct advice. A much clearer and completely unambiguous statement is provided by the following version of the First Law of Advice:

> The correct advice to give
> is the advice that is desired.

A classic example of good advice was given to the mayor of Pittsburgh, Pennsylvania, in the fall of 1920. A major city highway, built on the side of a hill, began sliding, a piece at a time, onto railroad tracks below. With every heavy rain, more mud and parts of the road washed down, causing many of the railroad tracks to be unusable. Efforts to remove the mud to keep all tracks operable had been too slow following some of the more massive mud slides.

Several solutions were offered by local engineering and construction firms and by concerned citizens. One was to pave the entire side of the hill to prevent erosion. Another was to build a metal structure to support the road and protect the tracks. All the suggestions would have been quite expensive, causing thoughtful people to wonder *how* or *if* the city could pay for the necessary work. Furthermore, no one knew if any of the proposed ideas would solve the problem.

AN EXPERT IS CALLED

The mayor knew he needed good advice. Ultimately, he hired G. W. Goethals as a consultant. Goethals had served as chief engineer for the Panama Canal and had acquired considerable experience with landslides. His expertise was evident not only in his experience, but also in his consulting fee of $1,000 per day—an extraordinary sum at that time.

After only one day of study, Goethals was prepared with his advice and with his bill. His advice to the city was simple: "Let it slide."

The opposition party and one of the newspapers made sport of the city administration for paying so much for this advice. The mayor rightly argued that it was a small price to pay to learn that none of the more expensive proposals would work. The mayor chose to follow this most economical advice and permitted the hill to slide.

Whether technically right or wrong, the consultant's advice was the desired advice. It did not involve construction expenses. Furthermore, any other solution would have been under constant attack from engineering and construction firms whose "solutions" had been rejected in favor of the winning contractor. The desired advice was clearly the correct advice.

This classic example also stands up well in terms of the Second Law of Advice:

> The desired advice is revealed
> by the structure of the organization,
> not by the structure of the technology.

And it also agrees with the Third Law of Advice:

> Simple advice is the best advice.

Simple advice is best.

Another classic example of advice that obeys all three laws was the advice given to the vice president of a petroleum company during the 1920s. The company had discovered a major oil deposit of high quality that could be refined economically into gasoline and other products. There was only one problem. The resultant gasoline had a greenish tint, which the refinery had been unable to remove. Because all gasoline at that time was clear, like water, the marketing group believed there would be considerable customer resistance to an impure-looking gasoline.

The production manager submitted his proposal for solving the problem. It called for modernization of the refinery. The company's chief chemist objected on the grounds that there was no proof that refinery modifications would result in a better product. Removing the greenish tint was a difficult chemical problem that had defied every attempt at solution. The chief chemist, therefore, recommended an expanded research program.

FOR ADVICE, GO OUTSIDE

Rather than adopting either solution, the vice president wisely turned to an outside consultant for advice. The consultant was a chemical engineer of good reputation in academic circles, who had consulted before in the petroleum industry.

He listened to the proposal of the chief chemist and then to that of the production manager. He talked to engineers and managers at the refinery and to chemists in the laboratory. Then he returned to the university for further study. If he were to recommend more experimental work, the chief chemist would be pleased. On the other hand, a recommendation to modernize the refinery was the desired advice of the production manager.

The important thing for the consultant, however, was to determine what advice was desired by the vice president. The vice president did not want to be responsible for choosing either of the proposals already presented. He wanted to avoid responsibility for any decision that would appear to favor either of his subordinates. If such a decision had to be made, it would be best to attribute the decision to an outsider. This, the consultant discerned, was the real reason why he had been hired. Even better for the vice president would be a totally different solution that played no favorites.

After several weeks of additional work, the consultant was ready with a uniquely neutral recommendation—one that required neither research work nor modernization of the refinery. His advice to the vice president was simple: "Advertise the color."

The marketing success of the greenish gasoline and the fact that most gasoline is now artificially colored demonstrate once again that advice found by studying the structure of the organization—not the structure of technology—is the desired advice. It also substantiates the Third Law. Indeed, simple advice is the best advice.

COLLECTIVE WISDOM

But what if the advisor is more simple than the problem? Then, simple advice will not be found.

This is a common dilemma, especially in large organizations where problems may involve a multiplicity of elements, each one of which is under the jurisdiction of a different group. Not only is responsibility diffuse, but so is understanding.

In such cases, the wisdom of a single individual is not enough. The proper response is provided by the Fourth Law of Advice, which deals with collective wisdom:

When in doubt, form a task force.

Task forces permit the skills from many areas to be pulled together temporarily to solve a problem. After the problem is solved, the members can return to their previous assignments. Task forces have become so popular that they are the standard mode of operation in many fast-moving organizations.

The reason for their popularity is clear. They offer something for everyone.

Each member of the task force is flattered to be identified as the most knowledgeable person in a given field—to have his expertise recognized and his advice sought. For the leader of the task force, it is an opportunity to address a big issue. But, of course, the greatest benefit of all accrues to the one who initiates the task force. If the advice is followed and the outcome successful, he can take full credit. If the outcome is not desirable, he can retreat behind the shield of having sought advice from the "best minds" in the field. Who could be faulted for that?

Whether the study is successful or not, the task-force initiator is recognized as a person of perception, a person capable of spotting important problems.

The task-force leader's most important function—aside from arranging for meeting times and places, coffee breaks, and transportation—is wordsmithing. With three other members on the task force, he will have to write the final report so as to reflect the four divergent opinions without appearing to equivocate. With ten or more members, wordsmithing is even more important.

Representing two or more views as one, or coming down with both feet planted firmly on three or more spots, is an art. Poor command of this art by a task-force leader may result in lingering doubt and necessitate action, as commanded by the Corollary to the Fourth Law of Advice:

When doubt persists, establish a committee.

The Consultant's Law

B ehemoth Insurance Corporation had developed one of the finest information technology (IT, pronounced "eye tee") departments in the business. Its improved profit position and renewed growth were due largely to its cost-effective use of information-processing methods.

Key to this development was Ed Vise. He had come directly from college to Behemoth, where he had matured as a systems analyst and programmer. No one in the IT department had a better understanding of the idiosyncrasies of the computer hardware and software systems. Nor had anyone at the center made a greater effort to learn about the insurance business. When an innovative solution was needed, the director of the department invariably turned to Ed for help.

Thus, while Ed managed a small group in the department, his primary role had evolved to that of chief advisor, or consultant, to the director. It was a satisfying position, and Ed was rewarded with the highest salary next to that of the director himself. Nevertheless, he became dissatisfied.

Ed longed for a more challenging and rewarding position. There was little chance that he would be made director of the department any time soon, and there was no other position for him in the company. His salary was good, but it did not reflect the millions of dollars the company had earned as a result of his unique contributions.

Putt's Law and the Successful Technocrat. By Archibald Putt
Copyright © 2006 the Institute of Electrical and Electronics Engineers

GOING IT ALONE

Not surprisingly, Ed Vise began thinking of becoming an independent consultant. He could supply his experience and skills to insurance companies throughout the industry. Because Ed's unique skills were largely unknown outside of Behemoth, he decided to charge his clients a low daily rate, but with a bonus for early completion of assigned tasks.

His first job was very successful. By using the same techniques he had developed at Behemoth, he was able to solve the problem given to him in about half the estimated time. He was rewarded with a large bonus and a letter of thanks that was invaluable in securing his next consulting assignment.

For several years, Ed Vise was successful at giving advice to IT departments throughout the insurance industry. Gradually, however, things became more difficult. The solutions he provided one company tended to be learned by other companies through informal discussions or through presentations at professional meetings. Such presentations were typically made by IT managers, who felt no obligation to acknowledge the contributions of outside consultants. Such presentations not only reduced the areas in which Ed could apply his ideas, but they failed to give him much needed publicity.

As time passed, each new consulting job required more innovative ideas because his original ideas had become broadly adopted. Bonuses were increasingly rare. Even a man as competent as Ed could not repeatedly invent and implement on short schedules. More and more he was scratching for his next job while trying to complete the last one. Finally, he began looking for a secure job in a large company, where pay and productivity were more loosely coupled.

A FORMULA FOR SUCCESS

Ed's major problem as a consultant was in paying too much attention to technology and not enough attention to the hierarchiology of consulting. If he needed to be convinced of how important this is, he had only to observe the fortunes of Harvey

Goodfellow, who had worked for him at Behemoth Insurance and who had also left to become a consultant.

Harvey was a good worker, but he lacked innovative capability. Under Ed Vise's management and guidance at Behemoth, he had done very well. Ed determined how the problems should be solved, and Harvey carried out the solutions. When Ed left, Harvey's performance dropped noticeably. Without Ed's innovative leadership, Harvey's solutions to problems became circuitous. No longer could he be counted on to get his assignments completed promptly.

The basis for his decline in performance was not understood by his new manager, and Harvey himself did not appreciate what had happened. He felt he was working as hard as ever but his new manager failed to appreciate his efforts. Within a year after Ed left, Harvey Goodfellow also left Behemoth to become a consultant.

His first customer was the same company that first hired Ed. Harvey followed through on some of Ed's suggestions that had not yet been implemented and then returned frequently to discuss their plans and problems. Harvey had shown himself to be competent by completing work outlined by Ed. He was an interesting talker and always congenial. Because he was unable to develop innovative solutions himself, he could discuss the ideas of the members of the IT department at length without getting bored.

He never embarrassed them by his brilliance. Frequently, he was invited to finish a day of consulting at a round of golf with the director of the IT department and the company's executive vice president. Eventually, he was placed on a retainer to consult a minimum of two days a month for the company.

Good references from this first consulting job led to new jobs in which a similar pattern developed. Within two years, Harvey Goodfellow had acquired eleven regular customers and was forced to advise his clients that he was now too busy to do any computer programming himself. This, however, only served to increase his value.

Harvey was now regarded as an expert IT consultant and could command a top price. Each of his clients had the benefit of all of the information he had gleaned from his other clients. As these continued to increase in number, the information he could share increased almost in proportion. He was careful not to give the identity of his sources lest he reveal proprietary information; and to better protect the identity of his sources, he frequently felt obliged to claim borrowed ideas as his own.

Harvey's business was booming just when Ed's was in steep decline. After several unsuccessful attempts to find a corporate job, Ed accepted Harvey's offer of a job. With Ed working for him, Harvey could now look forward to providing clients with innovative solutions. He believed, perhaps erroneously, that such a capability would bring additional business to the new firm of Goodfellow Associates.

Harvey had no formal training in the hierarchiology of consulting, yet he managed his career as effectively as if he had been a student of the Consultant's Law:

The value to a consultant of each discussion
is proportional to the information he receives,
independent of any information he may give in return,

and its corollary:

A successful consultant never gives
as much information to his clients
as he gets in return.

The truth of this law and its corollary are affirmed by Harvey's success and Ed's failure. It can also be derived through simple mathematical relationships.

THE MATHEMATICAL DERIVATION

The value to a customer of a discussion with a consultant (V_c) is equal to the information given by the consultant (I_g) times the price per unit (P_u) that the customer would be willing to pay for the information. Stated mathematically, this becomes

$$V_c = (P_u \times I_g)$$

The value of the same discussion to the consultant (V_o) can also be represented mathematically by

$$V_o = (P_u \times I_r) - (P_u \times I_g) + F$$

where ($P_u \times I_r$) is the value of the information received by the consultant, ($P_u \times I_g$) is the value of the information that he gives in return, and F is the fee paid to the consultant in dollars. Assuming that the customer pays a fee equal to the value of the advice he receives (an interesting, even if naïve, assumption), then $F = (P_u \times I_g)$ and the value of the discussion to the consultant becomes

$$V_o = (P_u \times I_r) - (P_u \times I_g) + (P_u \times I_g)$$

or simply

$$V_o = (P_u \times I_r)$$

This equation states that the value to a consultant of each discussion is proportional to the information he receives. Therefore, as stated in the Consultant's Law, it is completely independent of any information he may give in return. Even if one assumes the fee paid to him is not exactly equal to the value of advice given, the analysis produces a very similar result.

CODA

Successful consultants soon learn that, when it comes to information and advice, it is more important to receive than to give.

The failure of most technical consultants can be traced directly to their mistaken presumption that the function of a consultant is to give information and advice. In reality, a consultant's job is just the reverse.

Five Laws of Decision Making

Successful managers know they are measured on their ability to make decisions and, whether right or wrong, take action. Thus, they readily grasp the significance of the First Law of Decision Making:

Managers make decisions.

Nonmanagers, by contrast, seldom concern themselves with this law. They may, therefore, commit the fatal error of making all decisions themselves. This is particularly true in technical hierarchies, where low-level technologists continually make decisions in their areas of expertise. Finding so few issues that the manager is qualified to decide, they are likely to forget to bring any issues to him at all. This was one of Roger Proofsworthy's many problems at the Ultima Corporation. He performed all of his assignments with so much perfection that his accomplishments went unnoticed. Thus, he never progressed beyond the first level in the hierarchy and finally left in disgust.

The final blow to Proofsworthy at Ultima was management's rejection of his best project proposal. It was intended to provide Ultima with a continuing stream of new products. Proofsworthy called the project HOPE, an acronym for Highly Original Product Exploration.

Putt's Law and the Successful Technocrat. By Archibald Putt
Copyright © 2006 the Institute of Electrical and Electronics Engineers

Managers make decisions.

Management was initially interested, and Proofsworthy was asked to make numerous presentations. At first, the questions were quite limited in scope. But as interest increased, they were broadened to include competition, patents, manufacturing, marketing, and finance. Each time, Proofsworthy was ready. No matter what the topic, he had already considered it and had determined a proper course of action.

If management had accepted Proofsworthy's recommendation to initiate project HOPE, Ultima's future exploratory activities would have been fully determined. No additional decisions would be needed by management. The excessive perfection of Proofsworthy thus threatened to deprive management of its primary function.

The members of management reacted in the only way they could. They rejected the project. Thus, they assured themselves of making many more decisions.

One of their first decisions, after Proofsworthy left Ultima, was to assign a technologist, named Lisa Rightway, to look again into new product explorations. After careful study of Proofsworthy's proposal, Rightway made an "entirely new" proposal called MOPE, an acronym for Management Originated Product Exploration.

It was immediately well received.

With the help of Proofsworthy's earlier study, Rightway was able to solve most problems associated with the proposal. However, she was careful to reserve several items specifically for resolution and decision by management. With their involvement in the project thus assured, management eagerly supported it. Even the chairman of the board took delight in selecting the color for the walls of the laboratory in which project MOPE would be initiated.

Although deciding the wall color may sound trivial, it was one of the few topics about which all members of management had knowledge. All felt qualified to express a view, and did— until the *final* and *correct* view was given. It was correct not only because it was given by the chairman of the board, but also because it was expressed by a man with full command of the Second Law of Decision Making:

Any decision is better than no decision,

and also the Third Law:

> A decision is judged
> by the conviction with which it is uttered.

THE ANALYTICAL APPROACH

Decision making in a technical hierarchy is similar to that found elsewhere, except for the greater use of analytical decision-making methods. An example of this is provided by the case of Xavier Y. Ziegler, the newly appointed director of advanced development of the Solid Status Electronics Company. His promotion to director was based partly on his skill in using cost–benefit analyses.

Shortly after his promotion, some scientists and engineers proposed a new product. The idea was exciting, but it would require considerable development effort. X. Y. Ziegler was delighted with the opportunity to demonstrate his analytical decision-making skills. Within a week, he had assembled the talent required to evaluate the proposal. It included specialists from research, manufacturing, marketing, and finance. Ziegler worked on the entire study, paying particular attention to the cost–benefit analysis.

When the study was completed, a special meeting of the corporate officers was called. Ziegler began the presentation with a simplified version of the technical evaluation, which was nevertheless beyond the grasp of many attendees. The president and two of the vice presidents excused themselves for important phone calls. Several of the other attendees began doodling or whispering among themselves.

MATTERS WORSEN

X. Y. Ziegler tried to skip over material in order to reach his conclusions more quickly, but this was not possible. The technical staff members, who had been invited to attend, felt obliged to demonstrate their knowledge. They asked for clarification of many points and challenged some of the assumptions. Several of their questions revealed technical ignorance to Ziegler, but not to the corporate executives who were merely bored. Ziegler was becoming concerned that even the most inane questions were raising doubts about the validity of the assessment. As the discussions dragged on, the director of manufacturing excused himself in a coughing spell. Several others went out for a smoke or headed to the rest rooms.

It had not occurred to Ziegler before, but leaders in a technical hierarchy become accustomed to having all the technical decisions made by specialists at lower levels. Their own analytical skills atrophy. The selection of Ziegler for a midmanagement job was not intended to help upper management gain a better understanding of technical issues. Just the reverse was true. His job was to interpret technology so that upper management would not have to concern itself with such intellectually difficult problems.

Above Ziegler's level, all decisions regarding the technical direction of the organization were to be made without consideration of the technical issues. Ziegler was learning, through this presentation, about an important law that may be unique to technical hierarchies—the Fourth Law of Decision Making:

Technical analyses have no value
above the midmanagement level.

This law is illustrated in the figure on the following page. The value of technical analyses is highest among support and staff personnel who have the time and expertise to understand them. The value drops rapidly as one rises through the management levels, reaching zero between the department-director and vice-presidential levels. At all higher levels, it has a negative value.

A RESOUNDING NO!

Ziegler finally got to the summary and cost–benefit issues. The gist was that the return on investment for the new product, even if successfully developed, would not be as good as for the current products. Furthermore, it would require new facilities and large expenditures that would increase the company's financial

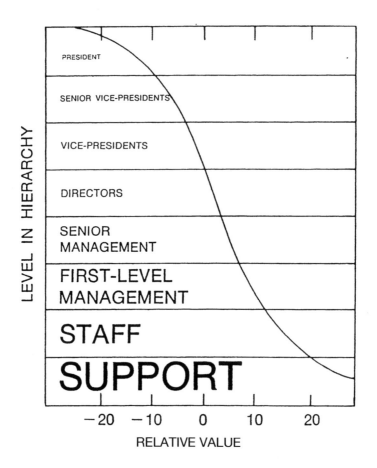

Value of Technical Analyses.

risk. Ziegler then showed that a relatively small additional expenditure on the present products would provide the same increase in revenue and even greater profits. It would also involve very little risk for the company.

Ziegler ended his presentation with a recommendation not to initiate the proposed product program. The attendees were stunned.

The program, if initiated, would have resulted in a substantial increase in Ziegler's own responsibilities. For the sales manager, it would have meant a larger sales force and new areas to cover. The director of manufacturing would have needed a larger factory. And the broadening of the product line would have benefited all through the enhanced status of the Solid Status Electronics Company.

Discussion of the presentation began hesitantly. The analysis appeared to be correct, but no one said so. A number of meaningless questions were asked and there were uncomfortable moments of silence. Finally, the director of product development sensed the feeling of the group and spoke up.

"The numbers are okay," he said, "but that's all it is—numbers. This company didn't get where it is through numbers. It got here through imagination, hard work, and taking risks. People smart enough to come up with ideas like this proposal can also come up with better ways to implement it than has been assumed by our new director of advanced development. Put those individuals into an aggressive product development group like mine and you can throw out all those numbers."

The other corporate officers agreed. The analysis indicated there was risk, but it did not rule out the *possibility* of success. Management, after all, was paid to take risks—and in this case, they would. The program and several of Ziegler's best people were transferred out of his area and placed under the director of product development. The program was initiated with full support of all officers of the corporation.

X. Y. Ziegler's failure to find support for his recommendation resulted from his failure to consider the costs and benefits to the decision makers. The question of *who* benefits is all important in any practical cost–benefit analysis. In making decisions, decision makers are generally concerned primarily with the costs and benefits to themselves. This is clearly stated in the Fifth Law of Decision Making:

Decisions are justified by
benefits to the organization;
they are made by considering
benefits to the decision makers.

In time, a number of technical problems arose that had been
predicted by Ziegler. To solve these, Ziegler and his remaining
people were placed under the supervision of the director of
product development. The president of Solid Status Electronics
simultaneously announced the elimination of Ziegler's previous
position of director of advanced development. That function, he
explained, was now contained within the product development
department.

Laws of Communication

S ir Charles Snow shook the intellectual and scientific communities in the 1950s and 1960s by his articles, books, and lectures on the "two cultures." His thesis was that the intellectual community, which historically provides leadership in political and cultural change, was now divided into two camps: the *literary intellectuals* and the *scientific intellectuals*. The failure of these two cultures to communicate with each other, Snow suggested, could be dangerous for the future of civilization.

C. P. Snow cites the frustration and puzzlement expressed to him by the great mathematician G. H. Hardy in the 1930s when he asked, "Have you noticed how the word 'intellectual' is used nowadays? There seems to be a new definition which certainly doesn't include Rutherford or Eddington or Dirac or Adrian or me. It does seem rather odd, don't y'know."

It did seem rather odd to Snow, who himself was provoked once or twice in the presence of literary intellectuals to ask how many could describe the Second Law of Thermodynamics. The response, he notes, was always quite negative. Yet the question being asked was the scientific equivalent of, "Have you read a work of Shakespeare's?"

Indeed, the world had come a long way in the two centuries since men such as Thomas Jefferson and Benjamin Franklin were the political, literary, and scientific leaders of their times.

Putt's Law and the Successful Technocrat. By Archibald Putt
Copyright © 2006 the Institute of Electrical and Electronics Engineers

But what has happened during the many decades since Snow proclaimed his concern over *two* cultures?

A WORLD OF *N* CULTURES

The rapid development of technology with its multiplicity of specialties has now thrust society from a world of two cultures into a world of *N* cultures. The letter *N* denotes a large number that enumerates the fields of specialization into which modern technologists fall: microbiologists and biophysicists; solid state, nuclear, and applied physicists; organic, inorganic, and biochemists; electrical, mechanical, nuclear, and civil engineers; and so on. The list is nearly endless in technology, and even the humanities are subdividing into areas of specialization that are unable to communicate effectively with one another.

The impact on society is enormous. At the very time that technology is becoming the dominant force in our lives, technologists are no longer able to communicate effectively with each other, let alone with the rest of society.

Knowledgeable technologists warn society about the dangers of nuclear power plants. Other experts, of equal stature, advise us that the dangers of nuclear energy are less than those of other sources of power.

Whom can one believe? How can nontechnically trained congressmen be expected to make correct decisions, when the technologists cannot agree among themselves?

BENEFITTING FROM CONFUSION

This state of confusion is not good for society, but it is good for ambitious technocrats and politicians who learn how to benefit from confusion. To be heard, they need only take strong positions. To defend their positions they can quote from others with the same views. With so much confusion, any strong position (right or wrong) will be welcome. This is not only true at the interface between technologists and nontechnologists, but it is also increasingly true at the interfaces among the various specialties within the technical community itself.

In this environment, ambitious technocrats and politicians can support their desires to rise in the system by understanding and heeding the First Law of Communication:

The purpose of communication
is to advance the communicator,

as well as the Second Law:

The information conveyed
is less important than the impression.

R. U. Tooley had spent the last ten years as custodian. During those years, he had made many suggestions. But because he did not understand the laws of communication, few of his suggestions were followed. For example, he had urged the company to replace its inefficient heating plant to save money on fuel and maintenance—but to no avail. Finally, a younger man, who was hired to work with Tooley, made the same suggestion. He also intimated that Tooley had failed to push vigorously for the replacement because he wanted to increase the number of maintenance men and become a manager.

The younger man's fresh perspective resulted in the installation of a new heating plant. Within a year, he had also become Tooley's boss. His communication was clearly more effective than Tooley's.

Roger Proofsworthy, that unsung hero of the technical hierarchy, also failed to obey the laws of communication. Thus, he failed to get support for his Project HOPE (Highly Original Product Exploration). Yet the same proposal was quickly supported by management when another engineer presented it as MOPE (Management Originated Product Exploration). Proofsworthy and Tooley not only failed to obey the First and Second Laws of Communication, but they also were oblivious to the Corollary to the Second Law:

It is not what you say, but how you say it.

This corollary is well observed by most politicians who frequently do not follow the advice for the timid, "When in doubt, mumble." Instead, they speak with their greatest force and conviction on issues over which they have the greatest doubt. This usually works well, but it is hazardous.

For technocrats, this hazard is much reduced. Each technical field is blessed with specialized terms understood only by experts—and sometimes not even by them. Thus, a skillful technocrat may be able to make many forceful statements without any fear of being understood.

Laws of the Information Age

Following a genetic change that gave *Homo sapiens* spoken-language capability, according to one theory, an estimated 50,000 years passed before a written language was developed. Within another 5,000 years, *Homo sapiens* had created versatile printing presses; and in less than 500 more years, by the end of the nineteenth century, they had developed the telegraph, telephone, phonograph, electric lighting, calculating machines, and punched card tabulating equipment.

The Information Age was born during the next 50 years with the introduction of radio, television, digital circuitry, transistors, fax machines, electronic computers, and Claude Shannon's mathematical theory of communication. The invention of transistors in 1947 and their incorporation in integrated circuits a dozen years later are particularly noteworthy. Dramatic progress in integrated circuits during the last half of the twentieth century defined the shape and scope of the Information Age.

MOORE'S LAW

In a paper published in *Electronics* in April 1965, Gordon Moore discussed the future of electronics. Among his predictions for integrated circuits was that the number of circuit components fabricated on a single silicon chip would double each year, reaching 64,000 by 1975. At the time, no chips had been manufactured with more than 64 components. His prediction was quoted widely because it seemed so unlikely and because Gordon Moore was the highly respected director of R & D for Fairchild Semiconductors. Three years later, he became a co-founder of the Intel Corporation.

Moore's prediction fit the facts so well that people began referring to it as Moore's Law. It is still known as Moore's Law, even when adjusted for a doubling period that had increased from one to two years by the end of the twentieth century.

As the number of circuit components per chip increased, the manufacturing cost per circuit decreased. This reduced the cost to manufacture the central processing unit and other computer parts that consisted primarily of electronic circuits. Engineers assigned to improve these computer parts were lucky. Those assigned to develop printers, displays, and especially software were not so lucky. There was no obvious way for them to achieve similar improvements, and management typically saw no obvious reason to forgive them when they failed.

THE PUTT–BROOKS LAW

In the early 1960s, Frederick P. Brooks was assigned to lead the development of software for IBM's proposed new line of computers. When announced in April 1964 as the IBM System/360, this product line was unprecedented in scope. Over one thousand people were employed in software development alone. Despite this effort, much of the software was not completed on schedule.

In describing his problems in a book published in 1978, *The Mythical Man-Month: Essays on Software Engineering,* Brooks wryly proclaimed what he called Brooks' Law: "Adding manpower to a late software project only makes it later." It was a pithy and accurate observation. The time spent by senior project members to educate new members typically exceeds the time available for new members to do effective work.

In retrospect, Brooks' Law should never have been limited to software. It applies equally well to any high-tech development project. I have, therefore, revised the law, which I generously credit to Brooks by referring to it as the Putt–Brooks Law:

> Adding personnel to a late
> high-technology project
> only makes it later.

A somewhat obvious corollary to this law states:

> All high-technology developments
> take longer than estimated.

LESSONS FROM HISTORY

Despite the Putt–Brooks Law and its corollary, high-technology developments raced relentlessly forward during the twentieth century. How was this accomplished? Was it the irresistible challenge presented by Moore's Law? Was it the emotional high that results from solving difficult problems? Or was it some more complex phenomenon?

Historians tell us that human experience is too complex to be fully analyzed. Thus, we should learn from the experiences of those who have gone before us. Military academies have followed this advice for decades by making military history an important part of the curriculum. Business schools have also made extensive use of historical "case studies."

In justifying this approach, educators are likely to cite a quotation attributed to George Santayana, from his book, *The Life of Reason,* Vol. 1 (1905): "Those who cannot remember the past are condemned to repeat it." Among the many subsequent revisions to this quotation, I prefer one that I call the First Law of History:

> Those who fail to learn the lessons of history
> are doomed to repeat the mistakes of the past.

More erudite persons may also cite a less well-known, and anonymous quotation, which I call the Second Law of History:

Every time history repeats itself,
the price goes up.

Unfortunately, the lessons of history are frequently ambiguous and contradictory. Thus, in my experience, a few well-constructed rules, laws, and corollaries are more valuable than many volumes of history. My views on the study of history are well expressed by what I modestly call Putt's Law of History:

Those who learn the lessons of history
are doomed to know when they are
repeating the mistakes of the past.

INTERESTING TIMES

"May you live in interesting times" is an ancient Chinese curse that seems particularly relevant to the Information Age. The spectacular rate of progress in semiconductor technology, predicted by Moore's Law, has made possible dramatic advances in computers, which in turn have significantly altered almost all areas of human endeavor.

Life styles of people in developed countries throughout the world have been changed forever. People in less-developed countries are increasingly experiencing the effects of the global economy that new technologies have made possible. Adjusting to these changes is exciting for some and traumatic for others. It is good for some and bad for others. But change is inevitable for all.

The technical people who lead, or are swept up in, this tidal wave of technological progress have an enormous challenge. Their achievements will be measured against an ever advancing set of goals and expectations. To succeed, they must learn to function in a world governed by the Primary Law of the Information Age:

Change is the status quo.

Information Technology Laws

T
he public first became aware of computers when Remington Rand's UNIVAC I was featured on the CBS election-night television program in November 1952. With only 5% of the votes tallied, UNIVAC I correctly predicted that Dwight D. Eisenhower would win the presidential election by a landslide over Adlai Stevenson, who had been leading in the polls.

So unexpected was this outcome, that the CBS program managers withheld the computer's prediction until they could check it against later information. Somewhat sheepishly the next day, they confessed to their actions. The computer had been right from the beginning.

THE ORACLE

The performance of UNIVAC I in 1952 has been followed by many awe-inspiring computer successes, such as landing men on the moon, beating the world's best chess player, deciphering the human genetic code, and connecting to the World Wide Web to make products, services, and information available throughout the world.

The computer has become the Oracle of the Information Age.

Putt's Law and the Successful Technocrat. By Archibald Putt
Copyright © 2006 the Institute of Electrical and Electronics Engineers

Oracle of the Information Age.

It was once said that marriages were made in heaven, but nowadays, it is more likely that they have been arranged by a computerized dating service. Coins minted by the United States government say, "In God We Trust," but the operations of government and industry are entrusted to computers.

People wishing to take action to solve a societal problem are well advised to get the problem formatted on a computer so the desired action can be shown as a computer solution. Even problems capable of simple solution may be solved *more convincingly* by a computer. As stated by the Law of Computer Users:

Actions are best justified
by computer solutions.

Despite the awe in which large computers are held, they have become favorite scapegoats. Organizations and individuals routinely blame computer error when things go wrong. But this does not diminish the reverence in which they are held.

When one says, "It is the fault of the computer," the impact is similar to that associated with the statement, "It is the will of God." By focusing blame on computers instead of on individuals, tensions between individuals and groups are reduced. This important social role of computers is the basis for a popular Rule of Computer Users:

Attribute successes to people
and problems to computers.

ACTUAL, ARTIFICIAL, OR VIRTUAL

Like oracles of old, computers often provide ambiguous information. Artificial intelligence and virtual reality, combined with the authority of computers, make it difficult to distinguish between what is actual and what is virtual.

Consider the plight of a woman who found exactly what she wanted on a store's display shelf. When she took it to the checkout counter, the clerk was unable to make the sale. "It's not listed in the computer," the clerk advised, "so I can't print out a sales slip."

"Surely you can sell it," the woman insisted.

"No," replied the clerk, authoritatively, "it's not in the computer, so it doesn't exist and can't be sold."

"Can I just take it with me?"

"Of course not," the clerk responded with irritation. "It's got the store's security tag."

In this case, the information and authority of the computer outweighed the physical evidence of the object itself.

In another case, we empathize with a woman who was attempting to correct her airline reservation over the telephone. She had listened to endless music and commercials while selecting among the many options offered by the computerized response system. Finally, she found herself emotionally responding to a salesman before she realized that he was just another recorded voice under computer control.

We can also sense the elation of a defendant who seemed certain of being convicted until a witness testified that the digital photographs had been altered. What had appeared to be real to the jury now seemed to be virtual.

When faced with difficult distinctions between actual and virtual, successful technocrats have learned to use the Pragmatic Rule of Reality:

When actual and virtual reality are blurred,
choose the one that serves you best,
and say it with conviction.

GOING ONLINE

At the beginning of 1993, the World Wide Web had only 50 known users. After the Internet was made available for commercial purposes in 1995, the use of e-mail and the World Wide Web skyrocketed. Ten years later, an estimated one billion people throughout the world were making use of the Internet. In the United States, well over half the people were connected to the Internet, and half of these spent more than three hours per day online.

Already, some people were going online to obtain all their information and merchandise and to conduct most of their business and social life. They were truly caught in the Web. Somewhat surprisingly, studies revealed that technocrats who were caught in the Web were more likely to be successful than those who were not. As stated by Putt's Paradox:

The more firmly you are caught in the Web,
the faster you can outpace your competition.

The pace of business has quickened. Communications, which a few years earlier would have been sent by "snail mail," are now sent by e-mail. Responses are typically received the same day, often within minutes. Even people at lunch or on vacation can review their e-mail by cell phone and respond immediately.

There is no time for ambitious technocrats to look beyond the Web for information, and there is no time for serious reflection between communications. The results can be disastrous. Nevertheless, the Law of Internet Usage continues to be affirmed:

Failure to keep up with the Internet
leads to failure in the race for success.

THE JUNGLE

Early users of the Web were primarily scientists and engineers who shared technical information. People who shared social or religious views soon formed Web sites, as did men and women seeking marriage prospects. Web sites were also formed by politically active groups, including terrorist organizations. There are sites for people interested in cannibalism or group suicide, and sites offering pornography have long been among the most popular and financially rewarding. Indeed, there are Web sites that cater to every imaginable human desire. According to one Internet Truism:

If somewhere it is so,
it is more so on the Internet.

Even many popular Web sites engage in ethically questionable

Caught in the Web.

activities. One of the more common is placing adware or spyware in a Web-site visitor's hard drive to obtain personal information. Of greater concern is outright fraud and theft practiced intentionally by bogus Web sites and unintentionally by legitimate sites that are victimized by the crafty thieves and hooligans who pervade the Internet.

The popularity of the Internet has made it attractive to organizations that send spam. These unsolicited communications are now the major part of Internet traffic. They are inexpensive for senders but costly to service providers and time-consuming to users. In addition to selling products and services or swindling naïve people out of their savings, spam may contain malicious software programs, designed to disable the recipient's computer.

Viruses and worms on the Internet are estimated to cost users and service providers over $100 billion a year, primarily in lost productivity. Government agencies and large corporations are increasingly active in attempting to protect individuals and the nation's infrastructure from cyber attacks.

Aware of these and many other risks, some individuals and groups have resisted getting involved in the Web. Nevertheless, the Web continues to grow. Valuable information and services lure people in, and there are not-so-subtle pressures from airlines and other vendors that are reducing their operating costs by replacing person-to-person services with Web-based transactions. It is only a matter of time before everyone will be caught in the Web. Once inside, the Law of the Web is self-evident:

There is no escape
from the World Wide Web.

Vast quantities of information available on the Web make it possible for researchers and authors to create publications at rates never dreamed of before. Plagiarism is so rampant that it seems socially acceptable.

Technocrats can no longer survive without help from an online computer to collect and analyze information to support a position or fend off attacks from competitors. Simultaneously, their computers must be kept up-to-date with the latest hardware and software to protect against the onslaught of computers maliciously programmed to do them harm.

It is comforting to know that, even in the seemingly lawless realm of the Internet, all the laws, corollaries, and tools of the trade discussed in this book apply. It is less comforting to know that the law of the jungle has been replaced by the Law of the Internet:

> Survival on the Internet always requires
> better hardware and software than you have.

Part Four

ADVANCED TOPICS

Project Selection

The preceding chapters provide a formal structure on which a technologist can base the *management* of high-technology projects. But what about the *selection* of projects? Is there any methodology that can be followed to select projects that are likely to succeed?

There has been a paucity of good advice in this regard because the success or failure of a project is largely determined by technological factors that are not understood until after the project is well under way. The probability of success for advanced development projects when started is well under 50%, and there is no general method for predicting which ones are more likely to succeed.

SOME SELECTION METHODS

It is rumored that a Harvard graduate student became intrigued with this problem and investigated a technique in which random numbers were assigned to research projects. The success or failure of these projects was then predicted using a formula based on the numbers carried by horses in the win, place, and show columns at Belmont. News of the remarkable success of his method, as compared to the more heuristic methods used by research directors, was said to have been suppressed.

Putt's Law and the Successful Technocrat. By Archibald Putt
Copyright © 2006 the Institute of Electrical and Electronics Engineers

It occurred to me that if one good approach had been suppressed, there could be others worth considering that had also been suppressed. With this in mind, I interviewed research directors and other high-level managers of technology to ask them how potentially good projects could be distinguished from bad ones.

The most common answer I got from successful managers was, "Just ask me, and I'll tell you which projects are good." But when I queried any manager about the quality of advice I should expect from one of his colleagues, the response was usually noncommittal. So I turned my attention to more promising areas.

The stock market is particularly interesting because the problem of selecting a good stock is similar to the problem of selecting a good research project. Selecting stocks based on "fundamentals" is frequently not successful unless the "fundamentals" include "inside information." In technology, however, all the expletives uttered in God's name do little to make available any "inside information" that He may possess.

Because most stock market services do not have inside information either, and would not share it if they did, they base many of their recommendations on "technical factors" that rely heavily on curves of past performance projected into the future. Even this approach has limited value for stocks because of the lack of universal laws governing their performance.

THE KEY TO SUCCESS

This, I realized, was the key to success in technology. There is a universal law governing all progress in technology. It is known as the S-Curve Law:

All progress in technology follows an S-curve.

The S-curve is the solid line in the figure on the following page. Although it does not provide any insight into the value of projects before they are initiated, it can be used to determine when (and if) projects will become successful.

In the early part of the curve, a lot of time and effort is ex-

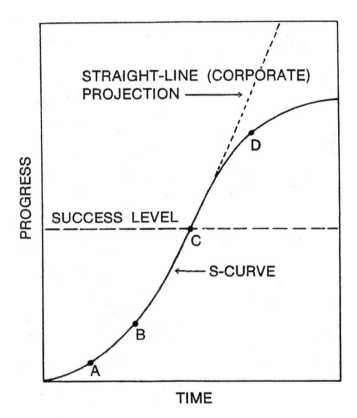

The S-Curve Law.

pended for very little progress. At this stage, there appears to be little chance of reaching the successful level indicated by the broken line in the figure. However, in projects that are destined to be successful, the pace of progress gradually quickens, as indicated at point *A* on the curve. By point *B*, the rate of increase of progress is quite noticeable. The rate will continue to accelerate further until a yet higher rate of progress is achieved well before the level of success is reached at point *C*.

Following the initial success of the project, progress usually continues at a rapid rate, leading to overoptimistic corporate, straight-line projections, as shown in the figure. Then a gradual leveling out occurs, which causes great trauma in the marketing and financial parts of the business. This, in turn, puts pressure on the technical groups because it is their job to solve the problems and get the technical progress back on the optimistic, straight-line projection.

Clearly, there is no reason to seek responsibility for a project in the early stages, when progress is slow and success uncertain. It is best to be given responsibility after point *B* (when progress is picking up) but before *C* (when the project has already been proclaimed a success).

There is temptation to remain with a project long after the point of success has been reached. This is technically referred to as "basking in the glory." It is, however, an ill-advised luxury. For once the rate of progress begins to level out at point *D*, it may be too late to avoid being blamed for failure to meet the straight-line projections.

Even worse than staying on a project too long is the mistake of stepping in to carry the ball once point *C* has been reached. By this point, further progress is simply *expected,* and it is more likely to be attributed to the technology than to the project manager.

Survival

There are times in any organization when advancement and promotions are not likely. The prudently ambitious technocrat then settles for survival. This requires a substantial adjustment in methods, because advancement demands risk, but, as stated in the Law of Survival,

> Survival is achieved through risk reduction.

The primary methods are the same as in any other hierarchy: conformity in dress, conformity in behavior, and, above all, conformity with the boss. These methods are learned early in life.

For example, consider a seven-year-old who was sent to school on a cloudy day dressed in her raincoat. Other children, who were not so dressed, teased her. She became unhappy and refused to wear her raincoat again even when it was raining.

A young man's first job after graduating from high school consisted of changing tires and balancing wheels for a local garage. He was soon able to change tires faster than his fellow workers, but they forced him to slow down—no one else wanted to work that hard. He began to waste time and dawdle between steps. At this slow pace, the work became dull and uninteresting, but he was afraid to work harder.

The seven-year-old and the young man had learned, even before the above examples, to adjust their life styles as commanded by the First Rule for Survival:

To get along, go along.

A local supermarket owner noted that his customers were unable to keep up with the rapid checkout process. By occasionally scanning the same item twice, his profits could be significantly increased. All clerks at the checkout counter soon learned that "errors" in favor of the store were part of the unwritten job description. To keep their jobs, they went along.

During congressional hearings on long-term care, two chiropractors, who had been convicted of fraud, testified about rampant abuses in the Medicaid program. They apparently went to the hearings "to complain about a system that is so bad that it virtually invites those acts" for which they had been convicted. While functioning in the system, however, they had found it expedient—and profitable—to go along.

Engineers in a large firm wrote a contract proposal for submission to the government in which the schedule and estimated costs were tight, but their management wanted a yet tighter schedule and lower cost estimate to be sure the firm won the contract. The engineers knew it would be impossible to do the engineering work on a shorter schedule and at lower costs, but reluctantly, they went along.

As a result, the government soon had one more contract running well over budget and behind schedule, thanks to the efforts of the engineers who joined the ranks of those who get along by going along.

PROTECTING YOUR POSITION

Conformity engendered by the First Rule for Survival ultimately leads to *stagnation*. This may be acceptable in some organizations, but it would never do in organizations devoted to innovation. Methods for avoiding *stagnation* are therefore in great demand.

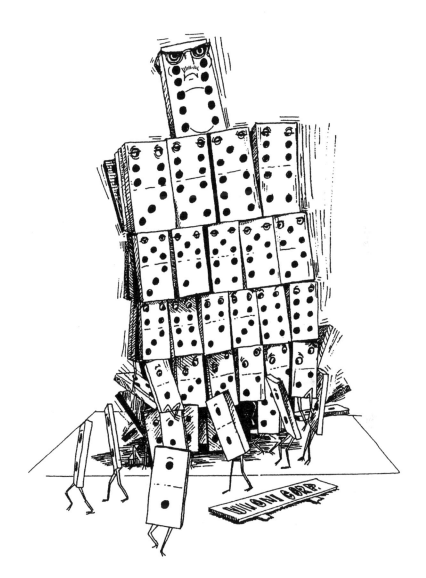

The domino firing strategy.

A variety of approaches are available, but the most common one is technically known as "regular trimming of the dead-wood." In this method, typically, 5 to 10% of the work force is fired every year—presumably from the bottom. Because it works so well in theory, it is frequently employed in practice. Ambitious technocrats must, therefore, be prepared to protect their own interests if such firing policies are adopted.

An excellent procedure for meeting arbitrarily established firing goals is to hire some people each year with the intent of firing them the next year. This avoids the unpleasantness of firing associates of long standing and helps to build a good manager–employee relationship among the "regulars."

A more sophisticated method is available that addresses one of the greatest concerns of many well-established technocrats. The concern is that they will be replaced by one of their energetic and ambitious subordinates. If management initiates a mandatory firing policy, one can avoid being replaced by someone from within the group by following the Second Rule for Survival:

To protect your position,
fire the fastest rising employees first.

By obeying this rule, managers avoid the common error of firing the lowest ranking employees first. Managers who make this error soon learn who was really doing the work. The remaining employees are much too senior to perform the mundane tasks that kept the project moving in the past. Progress diminishes and the manager comes under pressure to fire even more employees.

This is known as the "domino firing strategy." If the manager continues firing from the bottom, the process will continue until the manager is the last remaining member of the group. At this point, all activity in the project ceases, and top management can turn its attention to other, more pressing, problems.

Evaluating Ideas

Few things are as important to the success of a technocrat as the ability to evaluate ideas. Yet there are very few courses on this subject offered in colleges and universities.

A course of this type was initiated some years ago at a major school of technology. It has since been modified many times and copied by schools throughout the country. Students are typically given several different technical solutions to a problem and asked to evaluate them, or they are simply presented with the problem and asked to develop a solution. They are then graded on the quality of their analyses and on their ability to find the most cost-effective solutions. These are presumed to be the correct solutions.

Seldom, if ever, are students asked to consider the social or political implications of their ideas. Yet these implications are overwhelmingly important, as is revealed by the General Law Governing the Value of Ideas:

> The value of an idea
> depends less on its content
> than on the politics of its use.

WITHIN AN ORGANIZATION

This law is so well understood in most organizations that some readers may be surprised by the lack of understanding of it in technical organizations. This lack of understanding results from an educational system for science and engineering that is preoccupied with the sterile evaluation of technology. There is an underlying presumption that the world will automatically adopt the "best" engineering solution. How educators in technology can be so successful in promoting this erroneous concept and so unsuccessful in teaching useful concepts is not clear.

Despite this erroneous teaching, many engineers with bachelor degrees, and even a few with graduate degrees, manage to maintain some intuitive understanding of this law. They intuitively suspect that the value of an idea in an organization is diminished if the president of the organization dislikes it.

Although this intuitive understanding is useful, a more formal methodology for evaluating ideas is needed. Such a methodology has been developed through extensive research and analysis. The resulting Law Governing the Value of Ideas in an Organization asserts:

> The value of an idea in an organization
> is equal to the sum
> of the authority levels of its supporters.

If a person of authority opposes the idea, that person's authority level is entered into the summation as a negative number.

A simplistic determination of the authority levels in an organization assumes that the president has more authority than the vice presidents, that vice presidents outrank directors, and so on. In practice, however, authority levels are more finely divided and often deviate significantly from the organization chart. An ambitious technocrat must, therefore, continually monitor and fine-tune the actual levels of authority of all persons in the organization. By keeping a well-updated list on a computer, the technocrat will be able to calculate, almost instantly, the value of any new idea.

Such a list will also help in gaining support for one's own ideas. If you are low in the hierarchy, the support of people of authority is essential. Your search for the right group of supporters will be simplified by your updated list. Because highly ranked people are too busy being prestigious to generate ideas of their own, they will eagerly "admit" to having helped create yours. This is one of the best ways to acquire strong support.

SCHOLARLY PUBLICATIONS

The information and ideas contained within scholarly publications are essential to the advancement of technology. Thus, a means for measuring the value of each publication is needed.

Studies reveal that the prestige of the author or coauthors is the first measure of the value of a publication, quite independent of any new ideas the publication may contain. After writing a paper, it is thus worthwhile to find a prestigious person who is willing to become a coauthor.

Prestigious people are generally too proud to respond favorably to a crass invitation such as: "Please let me place your name on my paper as a coauthor. This will help me get recognition. It will also help you because you are much too busy to write original papers yourself."

It is far better to try a subtle approach, such as asking the person to comment on the paper. Then note that his comments have been so helpful that you believe he should be a coauthor. Few people of position can resist an offer like this.

It is not wise to load your paper with authors of equal or lesser rank just to raise its value. While the value of the paper will go up, the fact that this value must be shared among all the authors will outweigh the benefit to you. However, if you can get one or more colleagues to agree to include you on their papers as a coauthor—if you do the same for them—there will be a definite gain for all.

In addition to the prestige of the authors and the number of them on a paper, there is yet another way to measure a paper's value without trying to read and understand it. That method is to see how many subsequent papers refer back to it. This method has become very popular. It is based on the premise that authors refer only to papers that have been helpful to them or that they believe would be helpful to others. The use of these evaluation methods has led to the Law Governing the Value of Technical Publications:

> The value of a technical article, when first published,
> is proportional to the sum of the prestige of its authors,
> but its ultimate value is proportional to
> the number of references to it.

It should be noted that this law refers to the value of a paper to its authors, as opposed to any value it may also have to the technical community. This is because a technologist is advanced in the hierarchy by contributions of value to him. Contributions of value to technology serve only to advance technology.

FOR MATHEMATICIANS ONLY

Some readers will find the mathematical representation of the law more satisfying:

$$V_p = \frac{P_1 + P_2 + P_3 + \ldots + P_n}{1 + T} + N_r$$

In this equation V_p is the value of the publication to its authors, P_1 is the prestige of the first author, P_2 is the prestige of the second, and so on. N_r is the number of references to the paper in subsequent publications, and T is the time since its publication.

When the article is first published, $T = 0$; and since it could not yet have been referred to, N_r is also zero. Thus, initially, $V_p = P_1 + P_2 + P_3 + \ldots + P_n$, and the value of the paper is simply equal to the sum of the prestige of all of the authors. Five years after the article is published, $1 + T = 6$, and the importance of the authors' prestige is reduced to one-sixth of its original value. The value of the paper will then be more dependent on the number of literature references to it, N_r. After a long time, its value will be almost exclusively contained in N_r, as stated by the Law Governing the Value of Technical Publications.

GROUP APPROACH PAYS OFF

It is clear from this law and associated procedures for determining the value of publications that there is considerable stimulus for technologists to band together into "authoring groups." If we assume an authoring group of four persons in which each individual writes three papers per year and includes the other three individuals as coauthors, then each member of the group will have 12 papers to his credit at the end of one year, 24 at the end of two years, and 60 papers after only five years.

Even more important are the references to each other's papers in all subsequent papers. Assuming a one-year delay between writing a paper and having it cited, each of the 12 papers written the first year would be cited by 12 papers in the second year, making a total of 144 citations for each member of the group. By the end of five years, each member would have accumulated 1,440 citations! By increasing the group from four to eight persons, each member would after five years acquire part credit for 120 publications and 5,760 citations, and after ten years, 240 publications and over 26,000 citations. Even an Einstein could not match a record like that!

One problem with this scheme has been the rising cost of printing, coupled with the increasing size of the list of citations required at the end of each paper. After ten years, each paper would carry over 200 references to past publications even if no citations were given to persons outside the group.

Fortunately, this problem is easily overcome. After ten years, each of the eight authors would be so prestigious (with 240 publications and 26,000 citations to past publications) that the technical content of all future articles by the group would be incidental to their actual value.

The verbose articles of the past would no longer be needed. The authors could now concentrate on getting the concepts covered in as few words as possible. This would be greatly appreciated by younger technologists who would feel compelled to read all papers by such a prestigious group of authors. Eventually, each article could consist only of the title, the list of authors, a brief abstract, acknowledgments, and citations to past articles by the authoring group.

The ultimate in compact value would be a publication with no text and no abstract at all. Further shortening of the paper would cut too heavily into its value to be seriously considered.

Punishment and Reward

Is Putt's Ploy, as discussed in Part One, an aberration of an otherwise smoothly running organization? Or is it a logically consistent component of a well-tuned innovative organization?

Surprisingly, the latter may be the case. Many innovative organizations use a punishment–reward system that literally invites troubled employees to seek success through Putt's Ploy. This is illustrated by the upper-left-hand curve (for innovative hierarchies) in the figure on the following page, in which four different punishment–reward systems are illustrated.

Increasingly large positive values of y correspond to increasingly large rewards, and negative values of y correspond to punishment. Similarly, positive values of x correspond to success, and negative values of x to failure. In innovative systems, it can be seen that punishments occur only for small failures. Otherwise, relatively large rewards can be expected as the magnitude of either success or failure increases.

Such a reward system is justified because major innovations are not achieved without substantial risks and because success or failure of innovative projects cannot be predicted until long after they have been initiated. To encourage necessary risks, it is thus argued that rewards should be given to employees in proportion to the risks they take, whether these risks produce successes or failures. A number of experiences have revealed, however, that

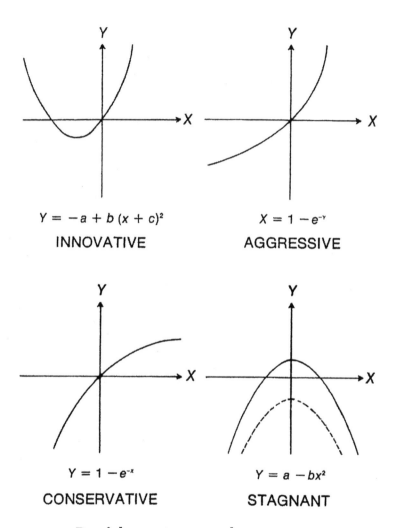

Punishment–reward systems.

some punishment for failure is necessary. These concepts are incorporated in the Rule for Managers of Innovation:

> Punish employees for small failures;
> reward them for big failures and all successes.

The other curves in the figure reveal substantially different reward systems for aggressive, conservative, and stagnant organizations. Readers are urged to ponder the impact of these other reward systems on employee performance. Theoretically inclined readers will want to make use of the equations in the figure as well as the curves. By proper adjustment of the parameters a, b, and c, these equations can be tuned to satisfy the requirements of almost any organization.

HIERARCHICAL AGING

Most organizations, in time, become more conservative. The weight of maturity and experience forces the ends of the punishment–reward curve down, particularly along the negative x-axis where failures occur. The technical name for this process is "hierarchical aging."

It begins as the innovative reward system is transformed, first into the aggressive one and then into the conservative one. These stages are illustrated in the figure. Finally, the heavy penalties, given even for small failures in conservative organizations, engender a sense of resentment toward those who succeed. A destructive undercurrent thwarts projects that might otherwise succeed. Persons responsible for successful innovations are attacked. Eventually, the punishment for success is as large as that for failure. Employees refuse to accept any risk at all as the organization succumbs to the Law of Stagnation:

> Stagnation of an organization occurs
> when the punishment for success
> is as large as for failure.

In advanced cases of hierarchical aging and organizational stagnation, no decisions are made that are not fully specified by "the book." Any attempt to deviate from the status quo is resisted. This is the familiar condition of government bureaucracies, the military, and educational institutions, about which so much has been written.

Two punishment–reward systems for stagnant hierarchies are illustrated in the figure. The lower, broken-line curve represents "malevolent stagnation," in which all activity, including inactivity, is punished. Enlightened stagnant bureaucracies have learned to offer small rewards for those who follow the conformist path. This is depicted by the solid curve.

Although the degree of stagnation is not altered, positive rewards do make life more pleasant for employees. Thus, stagnant organizations with positive rewards are described as being in a state of "beneficent stagnation."

THE NUMBER-ONE GOAL

Avoiding hierarchical aging must be the number-one goal of informed technical managers. They should stand firmly on the positive and negative x-axes, holding high the two ends of the reward-system curve. Skill is required to maintain the proper sag of the curve to provide the desired level of punishment for small failures.

Employees who understand the system can rise in it. Soon they too can experience the satisfaction of holding both ends of the curve.

Can Putt's Law Be Broken?

Tax cheats and bank robbers break laws of the government, and subatomic particles break Newton's Laws. So what about Putt's Law? Can it be broken?

"Why," you might ask, "should anyone want to break Putt's Law? Doesn't Putt's Law flush incompetent people out of lower levels into higher managerial positions, thus keeping lower levels filled with people qualified to do the work?"

This is true. It is one reason for the success of many technical organizations. But there is a downside. Putt's Law drives technical hierarchies relentlessly toward a state of competence inversion, as predicted by Putt's Corollary. Hierarchies with complete, or near complete, competence inversion can advance technologies along well-defined paths, but they cannot adapt to major external changes.

Therefore, aspiring technocrats need to know if and how Putt's Law can be broken. Histories of successful startup companies provide important insights.

STARTUP COMPANIES

Perhaps the most successful startup company of our times is the Microsoft Corporation. It began as an informal 50/50 partnership between Bill Gates and Paul Allen. These two nerdy college students had agreed to develop BASIC software in 1975 for the pioneering Altair personal computer. But before their partnership contract was signed, Gates revealed his now legendary negotiating prowess. He secured a 60% ownership for himself. Gates's initial investment was $910 and Allen's was $606. Less than 30 years later, Microsoft had a market value of more than $300 billion. Long before that, Bill Gates had become the youngest self-made billionaire in history, and generous stock options made multimillionaires of many other Microsoft employees. With Gates in charge, the company has never suffered a competence inversion.

Successful startup companies have many things in common. Typically, they are in the right business at the right time, they grow rapidly, and the geniuses who start them remain in charge. Technology projects become important so quickly that young innovators can accomplish their objectives only by leading teams of technical people. Thus, they are tricked into becoming managers! Some can be tricked again and again into ever-higher management levels.

Competence inversion in these companies is inhibited by many factors. The most important of all is rapid growth. The hierarchy of a large organization consists of many smaller internal hierarchies. During periods of rapid growth, new internal hierarchies must be added. So long as an organization grows rapidly enough, it will consist primarily of internal hierarchies too new to have reached competence inversion. My analysis reveals that an organization is unlikely to undergo complete competence inversion until at least half of its internal hierarchies have become inverted.

WHEN GROWTH SLOWS

Organizations can thus be saved from competence inversion simply by growing rapidly. But when growth stops, or is significantly slowed, the inevitable process of competence inversion in the internal hierarchies can bring down even the largest companies.

Ways to avoid this fate are eagerly sought. Restructuring so as to replace old internal hierarchies with new ones has been widely used. If this is properly done, each newly formed internal hierarchy will have little evidence of past slides into competence inversion and they will have to begin this process anew.

Other approaches used by large companies include mergers and acquisitions and offering early retirement incentives. The benefits of all these methods are limited, however, and generally support the validity of the Second Corollary to Putt's Law:

The impact of Putt's Law can be delayed,
but the Law itself is never broken.

Putt's Corollary to Murphy's Law

No discussion of the laws of innovative organizations would be complete without some mention of Murphy's Law. As revealed in this chapter, however, it is Putt's Corollary to Murphy's Law that is of most interest to the ambitious technocrat.

Murphy's Law itself was first pronounced in 1949 by an engineer named Edward A. Murphy, who was working on a U.S. Air Force project. He was obviously a man of great perception, for even though his law has never been proved mathematically, it has been verified experimentally again and again—not only in research and development, but in all activities in which modern technology is used. Thus, without the benefit of publication in scientific journals or discussions at meetings of learned societies, Murphy's Law has become known to technologists and nontechnologists alike:

Anything that can go wrong will go wrong.

This law appeals not only to our intellect but also to our sense of fair play, for it applies equally to all. During the first debate of the 1976 presidential campaign, for example, millions of dollars had been spent to insure uninterrupted television coverage. Backup units were on hand for *virtually* every piece of

Putt's Law and the Successful Technocrat. By Archibald Putt
Copyright © 2006 the Institute of Electrical and Electronics Engineers

electronic equipment. But then it happened. A simple thirty-five-cent electrolytic capacitor failed, and with it failed the audio portion of the telecast. The President and his challenger shifted from one foot to another in a twenty-seven minute "silent debate," while one hundred million viewers watched. The action of Murphy's Law had suddenly placed the candidates on the level of the viewers, wherein one thing that could go wrong did go wrong.

Who was to blame? Was it the person who designed the TV system for broadcasting the debates or the one who assembled it? Was it the manufacturer or distributor of the complex electronic equipment, or was it perhaps the worker who tested the capacitor at the end of the assembly line? The correct assignment of blame is difficult when complex technology is involved, but the Popular Corollary to Murphy's Law always applies:

> Anyone else who can be blamed should be blamed.

Just like Murphy's Law, this corollary appeals to our sense of fair play. It would not be fair if we, but not others, were blamed.

The pervasive use of computers has added yet another dimension to Murphy's Law. For these machines, if not flawlessly programmed and instructed, may work at lightning speed to produce more errors in one minute than could be made in a year by hundreds of hardworking, but misguided, mortals. The result of such computer effort is familiar to all and has resulted in the creation of a Computer Version of Murphy's Law:

> Anything that can go wrong
> will go wrong faster with computers.

A DIFFERENT APPROACH

Whether one cites the original version or the computer version, Murphy's Law is always a law of hopelessness and resignation. Blaming anyone else who can be blamed addresses this problem, but does not solve it.

What is needed is a fundamentally different approach to Murphy's Law. This is now available through a logically necessary and consistent corollary that has been developed following years of research. It is modestly known as Putt's Corollary to Murphy's Law:

If nothing can go wrong, it will go right.

But are there times when nothing can go wrong? The answer is, "Yes, there are such times. Nothing can go wrong anytime it has *already* gone right."

To be successful, technocrats must therefore work to be the *first* to know. They must have trusted informers. And they must subdivide the activities that report to them, so that they, alone, have all the facts. Only then can they *predict,* with confidence, when it is that nothing will go wrong.

Whenever Putt's Corollary to Murphy's Law applies, you will meet the truly successful technocrats. They orchestrate the event, report the action, and modestly refuse credit—except for the outcome. To join their ranks, you need only learn to do the same.

·

Part Five

PUTT'S CANON

Putt's Primer

Putt's Law:

> Technology is dominated by two types of people:
> those who understand what they do not manage
> and those who manage what they do not understand.

Putt's Corollary, also known as the First Corollary to Putt's Law:

> Every technical hierarchy, in time,
> develops a competence inversion.

First Law of Crises:

> Technical hierarchies abhor perfection.

Second Law of Crises:

> The maximum rate of promotion is achieved
> at a level of crisis only slightly less
> than that which results in dismissal.

Law of Failure:

> Innovative organizations abhor little failures
> but reward big ones.

Corollary to the Law of Failure:

> Failure to fail fully is a fool's folly.

Putt's Ploy:

> If you must fail, fail big.

First Law of Innovation:

> An innovated success is as good
> as a successful innovation.

Second Law of Innovation:

> The true measure of success
> in an innovative project
> is the size of management's reward.

Third Law of Innovation:

> Innovation may be the goal,
> but technology transfer is the business
> of technical hierarchies.

Law of Invention:

> It is more expensive to create inventions
> than to buy small companies that have them.

Rule of the Deal:

> When money deals are made,
> get a seat at the table.

Law of Leadership Potential:

> Those who see problems become whistleblowers;
> those who see opportunities become leaders.

Rule of Project Leadership:

> When leading a high-tech project,
> keep your eyes on the exit doors.

Method of Rational Exuberance:

> With each decision you make,
> publicly exude enthusiasm and optimism;
> privately calculate your chances for personal success.

Basic Putt

First Law of Innovation Management:

>   Management by objectives
>   is no better than the objectives.

Second Law of Innovation Management:

>   A manager cannot tell
>   if he is leading an innovative mob
>   or being chased by it.

Third Law of Innovation Management:

>   Stay in the pack until the objectives are clear.

Law of Insubordination:

>   Rejection of management objectives
>   is undesirable when you are wrong
>   and unforgivable when you are right.

First Law of Advice:

> The correct advice to give
> is the advice that is desired.

Second Law of Advice:

> The desired advice is revealed
> by the structure of the organization,
> not by the structure of the technology.

Third Law of Advice:

> Simple advice is the best advice.

Fourth Law of Advice:

> When in doubt, form a task force.

Corollary to the Fourth Law of Advice:

> When doubt persists, establish a committee.

Consultant's Law:

> The value to a consultant of each discussion
> is proportional to the information he receives,
> independent of any information he may give in return.

Corollary to the Consultant's Law:

> A successful consultant never gives
> as much information to his clients
> as he gets in return.

First Law of Decision Making:

>Managers make decisions.

Second Law of Decision Making:

>Any decision is better than no decision.

Third Law of Decision Making:

>A decision is judged
>by the conviction with which it is uttered.

Fourth Law of Decision Making:

>Technical analyses have no value
>above the midmanagement level.

Fifth Law of Decision Making:

>Decisions are justified by
>benefits to the organization;
>they are made by considering
>benefits to the decision makers.

First Law of Communication:

>The purpose of communication
>is to advance the communicator.

Second Law of Communication:

>The information conveyed
>is less important than the impression.

Corollary to the Second Law of Communication:

> It is not what you say, but how you say it.

Putt–Brooks Law:

> Adding personnel to a late
> high-technology project
> only makes it later.

Corollary to the Putt–Brooks Law:

> All high-technology developments
> take longer than estimated.

First Law of History:

> Those who fail to learn the lessons of history
> are doomed to repeat the mistakes of the past.

Second Law of History:

> Every time history repeats itself,
> the price goes up.

Putt's Law of History:

> Those who learn the lessons of history
> are doomed to know when they are
> repeating the mistakes of the past.

Primary Law of the Information Age:

> Change is the status quo.

Law of Computer Users:

> Actions are best justified
> by computer solutions.

Rule of Computer Users:

> Attribute successes to people
> and problems to computers.

Pragmatic Rule of Reality:

> When actual and virtual reality are blurred,
> choose the one that serves you best,
> and say it with conviction.

Putt's Paradox:

> The more firmly you are caught in the Web,
> the faster you can outpace your competition.

Law of Internet Usage:

> Failure to keep up with the Internet
> leads to failure in the race for success.

Internet Truism:

> If somewhere it is so,
> it is more so on the Internet.

Law of the Web:

> There is no escape
> from the World Wide Web.

Law of the Internet:

> Survival on the Internet always requires
> better hardware and software than you have.

Advanced Topics

S-Curve Law:

All progress in technology follows an S-curve.

Law of Survival:

Survival is achieved through risk reduction.

First Rule for Survival:

To get along, go along.

Second Rule for Survival:

To protect your position,
fire the fastest rising employees first.

Putt's Law and the Successful Technocrat. By Archibald Putt
Copyright © 2006 the Institute of Electrical and Electronics Engineers

General Law Governing the Value of Ideas:

> The value of an idea
> depends less on its content
> than on the politics of its use.

Law Governing the Value of Ideas in an Organization:

> The value of an idea in an organization
> is equal to the sum
> of the authority levels of its supporters.

Law Governing the Value of Technical Publications:

> The value of a technical article, when first published,
> is proportional to the sum of the prestige of its authors,
> but its ultimate value is proportional to
> the number of references to it.

Rule for Managers of Innovation:

> Punish employees for small failures;
> reward them for big failures and all successes.

Law of Stagnation:

> Stagnation of an organization occurs
> when the punishment for success
> is as large as for failure.

Second Corollary to Putt's Law:

> The impact of Putt's Law can be delayed,
> but the Law itself is never broken.

Murphy's Law:

> Anything that can go wrong will go wrong.

Popular Corollary to Murphy's Law:

> Anyone else who can be blamed should be blamed.

Computer Version of Murphy's Law:

> Anything that can go wrong
> will go wrong faster with computers.

Putt's Corollary to Murphy's Law:

> If nothing can go wrong, it will go right.

Printed in Poland
by Amazon Fulfillment
Poland Sp. z o.o., Wrocław
15 October 2020

1185a38c-7232-47ac-936b-1fb44a7851b7R01